Adrian Alfred Bischoff

Waste reduction in aquaculture by culturing detritivorous organisms

Adrian Alfred Bischoff

Waste reduction in aquaculture by culturing detritivorous organisms

Small steps towards sustainable aquaculture practices

Südwestdeutscher Verlag für Hochschulschriften

Impressum / Imprint
Bibliografische Information der Deutschen Nationalbibliothek: Die Deutsche Nationalbibliothek verzeichnet diese Publikation in der Deutschen Nationalbibliografie; detaillierte bibliografische Daten sind im Internet über http://dnb.d-nb.de abrufbar.
Alle in diesem Buch genannten Marken und Produktnamen unterliegen warenzeichen-, marken- oder patentrechtlichem Schutz bzw. sind Warenzeichen oder eingetragene Warenzeichen der jeweiligen Inhaber. Die Wiedergabe von Marken, Produktnamen, Gebrauchsnamen, Handelsnamen, Warenbezeichnungen u.s.w. in diesem Werk berechtigt auch ohne besondere Kennzeichnung nicht zu der Annahme, dass solche Namen im Sinne der Warenzeichen- und Markenschutzgesetzgebung als frei zu betrachten wären und daher von jedermann benutzt werden dürften.

Bibliographic information published by the Deutsche Nationalbibliothek: The Deutsche Nationalbibliothek lists this publication in the Deutsche Nationalbibliografie; detailed bibliographic data are available in the Internet at http://dnb.d-nb.de.
Any brand names and product names mentioned in this book are subject to trademark, brand or patent protection and are trademarks or registered trademarks of their respective holders. The use of brand names, product names, common names, trade names, product descriptions etc. even without a particular marking in this works is in no way to be construed to mean that such names may be regarded as unrestricted in respect of trademark and brand protection legislation and could thus be used by anyone.

Coverbild / Cover image: www.ingimage.com

Verlag / Publisher:
Südwestdeutscher Verlag für Hochschulschriften
ist ein Imprint der / is a trademark of
AV Akademikerverlag GmbH & Co. KG
Heinrich-Böcking-Str. 6-8, 66121 Saarbrücken, Deutschland / Germany
Email: info@svh-verlag.de

Herstellung: siehe letzte Seite /
Printed at: see last page
ISBN: 978-3-8381-3274-7

Zugl. / Approved by: Kiel, Universität, Diss., 2007

Copyright © 2012 AV Akademikerverlag GmbH & Co. KG
Alle Rechte vorbehalten. / All rights reserved. Saarbrücken 2012

Table of Contents

Chapter 1 — 7
1. Introduction — 8
1.1 General principles of aquaculture — 8
1.1.1 Production by environment — 9
1.1.2 Production systems utilised in aquaculture — 10
1.1.2.1 Ponds — 10
1.1.2.2 Tanks — 10
1.1.2.3 Cages — 11
1.2 Environmental impacts of / on aquaculture — 11
1.2.1 Aquatic pollution from aquaculture — 11
1.2.2 Pollution impacts on aquaculture — 12
1.3 Mono-, Poly- and Integrated aquaculture — 13
1.3.1 Monoculture — 13
1.3.2 Polyculture — 14
1.3.3 Integrated aquaculture — 14
1.4 Outline of this work — 15

Chapter 2 — 17
Abstract — 18
1. Introduction — 19
2. Material and Methods — 20
2.1 General description of the recirculating system — 20
2.1.1 System configuration of MARE I — 22
2.1.2 System configuration of MARE II — 22
2.2 Measurements and Methods — 23
2.2.1 Chemical parameters of the water — 23
2.2.2 Solid components — 23
2.2.3 Biomass determination — 25
3. Results — 26
3.1 Chemical parameters of the water — 26
3.2 Growth performance — 29
4. Discussion — 39
5. Conclusions — 47

Chapter 3	48
Abstract	49
1. Introduction	50
2. Material and Methods	55
2.1 *Nereis diversicolor*	55
2.2 Experimental set up	55
2.2.1 Survival	56
2.2.1. Is it possible to culture *N. diversicolor* on an exclusive diet of solid waste from fish culture?	56
2.2.1.1.1 Batch culture	56
2.2.1.1.2 Small scale recirculating system	57
2.2.1.1.3 Medium scale recirculating system	57
2.2.1.2 Impacts of sediment on survival	57
2.2.2 Growth	58
2.2.3 Consumption of solid waste by *N. diversicolor*	58
2.3 General experimental considerations	59
3. Results	60
3.1 Abiotic water parameters	60
3.2 Is it possible to culture *N. diversicolor* on an exclusive diet of solid waste from fish culture?	61
3.2.1 Dissolved inorganic nutrient concentrations	61
3.2.2 Survival of *N. diversicolor*	63
3.3 Which impact has the type of sediment on the survival of *N. diversicolor*?	66
3.4 Growth of *N. diversicolor*	66
3.5 Growth performance of *N. diversicolor*	71
3.6 Total organic matter contents of the sediment	72
4. Discussion	74
4.1 Survival of *N. diversicolor*	74
4.1.1 Is it possible to culture *N. diversicolor* on an exclusive diet of solid waste from fish culture?	74
4.1.2 Which impact has the type of sediment on the survival of *N. diversicolor*?	75
4.2 Growth of *N. diversicolor*	76
4.2.1 What influence has the type of sediment on the growth of *N. diversicolor*?	76

4.2.2 What is the optimum achievable growth under the applied conditions?	77
4.2.3 Are the applied conditions adequate to complete a lifecycle of *N. diversicolor*?	78
4.2.4 Is the total organic matter content of the sediment a reliable indicator for the consumption of solid waste by *N. diversicolor*?	80
5. Conclusions	81

Chapter 4	**83**
Abstract	84
1. Introduction	84
2. Material and Methods	86
3. Results	88
4. Discussion	95
5. Conclusions	100

Chapter 5	**102**
Abstract	103
1. Introduction	103
2. Material and Methods	104
2.1 Experimental set up	104
2.2 Sampling procedure	105
2.3 Prokaryote counts	106
2.4 Nitrification potential (Slurry assay)	106
2.5 Abiotic parameters	107
3. Results	107
3.1 Bacteria	108
3.2 Nitrification potential	110
3.2.1 Fine sediment	111
3.2.2 Coarse sediment	111
4. Discussion	112
4.1 Bacterial abundance	112
4.2 Taxonomic composition of nitrifying bacteria	112
4.3 Nitrification potential	113
5. Conclusions	114

Chapter 6 115

Abstract 116
1. Introduction 116
2. Material and Methods 117
2.1 Measurements of abiotic parameters 117
2.2 Analytical procedures 117
2.2.1 Dissolved inorganic nutrients 117
2.2.2 Water content of materials 118
2.2.3 Total organic matter content of materials 118
2.2.4 Energy content of materials 118
2.2.5 Carbon and nitrogen content of materials 118
2.2.6 Growth of *Crangon crangon* 118
2.2.7 Statistical analyses 119
2.3 Experimental set up and design 119
2.4 Experimental duration 120
3. Results 121
3.1 Abiotic parameters 121
3.2 Dissolved inorganic nutrients 121
3.3 Biochemical composition of applied food sources 122
3.4 Total organic matter content of the sediment 123
3.5 Survival of *C. crangon* 123
3.6 Growth of *C. crangon* 124
3.7 Biochemical composition of *C. crangon* 126
4. Discussion 129
5. Conclusions 134

Chapter 7 136

7.1 Is it possible to achieve a reduction of the solid waste load from aquaculture systems by the cultivation of detritivorous organisms? 137
7.2 What are the benefits of producing secondary organisms? 138
7.2.1 Increased consumption of supplied nutrients 138
7.2.2 Reduction of water exchange of recirculating aquaculture systems 138
7.2.3 Economical diversification of aquaculture endeavours 138
7.3 Which steps towards sustainability can be achieved? 138

7.4 Which criteria need to be fulfilled to integrate successfully
    detritivorous organisms into aquaculture? 139

References **140**

Chapter 1

General Introduction

Bischoff A.A.

1. Introduction

Fisheries play an important role in terms of global food production, with approx. 20% of human protein supply derived from aquatic habitats (Heise *et al.* 1996, probably on wet weight basis). Despite predictions of an endless supply of resources from the sea, the wild fishery harvest has stabilized over recent decades (Fig. 1).

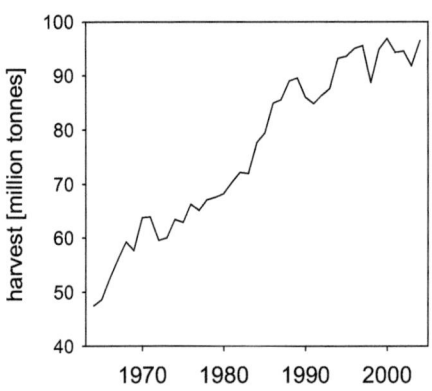

Fig. 1: Global fishery harvest over the last four decades according to FAO (2006a).

McVey *et al.* (2002) concluded that *'...the world human population has grown to the point where we can no longer expect to obtain additional protein from the sea without moving into the husbandry of the food species that are desired in the human marketplace'*. They further stated that *'...the capture fisheries have decimated many species of fish, crustaceans and molluscs leading to disruption of the natural balances in nature'*. Their final conclusion was that *'...new food from the sea for human consumption can only occur through aquaculture, just as it did for terrestrial systems through agriculture'*.

1.1 General principles of aquaculture

According to the Food and Agriculture Organization of the United Nations (FAO) Aquaculture is defined as *'...the farming of aquatic organisms including fish, molluscs, crustaceans and aquatic plants. Farming implies some sort of intervention in the rearing process to enhance production such as regular stocking, feeding, protection from predators, etc. Farming also implies individuals or corporate ownership of the stock being cultivated'* (Ottolenghi *et al.* 2004).

Aquaculture has been for long time the fastest growing sector within fisheries with constant positive growth rates during recent decades. It has shown annual growth rates of 9.2% during the last three decades. Total aquaculture production in 2004 amounted to more than 59 million tonnes wet weights (Fig. 2, FAO 2006a), which includes the overall production of fish (~28 million tonnes), molluscs (~13 million tonnes), plants (~14 million tonnes) and crustaceans (~4 million tonnes).

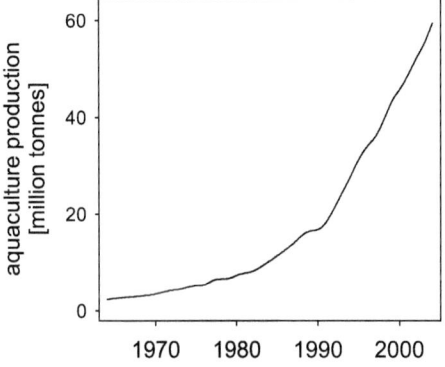

Fig. 2: Global aquaculture production over the last four decades according to the FAO (2006a). Production includes fish, molluscs, crustaceans and plants.

1.1.1 Production by environment

Aquaculture is conducted in various aquatic environments including fresh, brackish and marine waters. Freshwater production, which accounts for 43% of total aquaculture production, is dominated by cyprinids, mainly produced in earthen ponds of integrated systems in China and in south-east Asia (FAO 2006a). Mariculture, which is according to the FAO defined as aquaculture in brackish and marine waters, accounted for 57% of the total aquaculture production, or in absolute biomass for approx. 34 million tonnes in 2004. This value has to be subsequently divided into a larger part for marine production (approx. 30 million tonnes) and a smaller fraction for the production in brackish waters (> 3 million tonnes). This division in marine and brackish water aquaculture is mainly due to administrative reasons and overlaps much more in reality. Fig. 3 presents the ten most important families that were responsible for more than 45% of the total mariculture production in 2004. Aquatic plants such as macroalgae and seaweeds are excluded from this figure. These ten families include seven families of molluscs (Fig. 3: families 1 – 3, 5 -7 and 10), two families of fish (Fig. 3: families 4 and 9) and one family of crustaceans (Fig. 3: family 8).

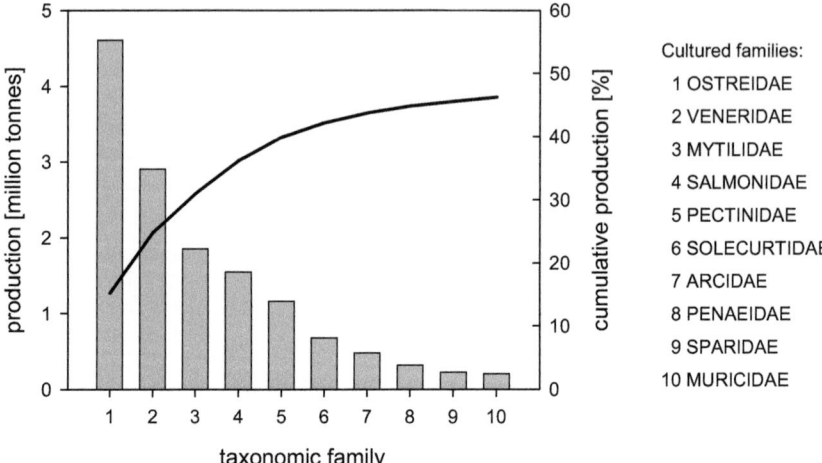

Fig. 3: Mariculture production 2004 according to FAO (2006a) – bars represent the ten most produced taxonomic families in mariculture (names are given on the right hand side). The line represents the cumulative production of the ten presented families.

1.1.2 Production systems utilised in aquaculture

Although particular facilities utilised in aquacultural will be described separately in the following section, normally more than one of each of these structures will be applied during the whole lifespan of cultured organisms.

1.1.2.1 Ponds

Ponds, which may be nothing more than a hole in the ground, are the oldest and most widely used structures in aquaculture. According to Lucas and Southgate (2003), their main requirements include a reliable water supply, relatively impermeable soils for construction, well-structured soils with good organic matter content to support pond ecosystems and gravity drainage.

1.1.2.2 Tanks

Tanks, similar to ponds, are commonly used in aquaculture. They are usually situated above ground and may be used in-, or outdoors. Tanks are used in a wide variety of size and shape, depending on the particular purpose they are used for. Tanks normally utilise a water supply (inlet) and a drainage system (outlet), with the function of the inlet being regulation of water exchange. The drainage of the systems water, including the removal of the solids that gather on the tank bottom (e.g. faeces, waste food), is regulated by the outlet.

1.1.2.3 Cages

Modern cages used in aquaculture are devices that float in the water reaching either the surface and include integral nets below the surface to confine cultured animals, or submerged under the water surface. Cages are regularly used for the grow-out phase of fish to reach their market size. Cages are open, allowing full water movement, and thereby removing dissolved and particulate nutrients originating from the cultivation of the fish.

1.2 Environmental impacts of / on aquaculture

1.2.1 Aquatic pollution from aquaculture

Numerous threats caused by aquaculture such as escapes from culturing non-indigenous species (Stickney 2002), genetic changes caused by the escape of cultivated fish into natural populations (Hershberger 2002), transfer of diseases (Stickney 2002), or the release of chemicals used for aquaculture such as therapeutants or antifoulant (Brügmann 1993; Alterman et al. 1994) are recognized. In the following section nutrient pollution caused by aquaculture will be addressed in more detail.

All of the cultured families presented in Fig. 3 excrete nutrient waste during their production. According to Schneider (2006), waste can be described as the difference between feed intake and weight gain, plus some other products. Non-retained nutrients are excreted as faecal loss in particulate form, or as non-faecal loss in dissolved form. This comprise mainly faecal loses and dissolved nutrient excretion from the cultivated animal as well as uneaten feed. Therefore, the culture of aquatic animals always produces waste in either one or both of the mentioned forms. The production of waste depends on a number of different factors such as species, animal size and stocking density, which combined determine the amount of applied food. Dosdat et al. (1996) showed that fishes like sea bass (*Dicentrarchus labrax*), sea bream (*Sparus aurata*), brown trout (*Salmo trutta*) and rainbow trout (*Oncorhynchus mykiss*) have ammonia excretion rates of 30 – 38% whereas turbot (*Scophthalmus maximus*) has a lower ammonia excretion rate of 20%. Results presented by Kim et al. (1998) for rainbow trout and Lupatsch et al. (2001) for sea bream were in agreement with these outcomes. The effect of animal size on nitrogen excretion rates was reported by Harris and Probyn (1996) for white steenbras (*Lithognathus lithoghnathus*) and showed increased endogenous ammonia excretion rates for smaller fish. Tatrai (1986) found a combined effect of temperature and fish body weight influencing the total nitrogen excretion for bream (*Abramis brama*). Lachner (1972) examined the effects of stocking density on nitrogen excretion. He reported that an increase in stocking density from 5 to 50 kg m^{-3} led to a 20-fold increase of ammonia excretion.

Besides the factors mentioned above, further aspects influencing waste production are the type and composition of the food supplied, the feeding regime and the experience of the workers. Ackefors and Enell (1994) as well as Cho and Bureau (2001) described improvements for reducing waste output through improving diet formulation and the strategies used during feeding. Results by Boujard et al. (2004) showed that an increase in dietary lipid level led to a significant decrease in voluntary feed intake, without affecting growth rates. They reported further that nitrogen excretion was related inversely to the dietary lipid levels; and by increasing the dietary lipid levels the nitrogen loss of fish produced was reduced. Peres and Oliva-Teles (2006) investigated the effect of dietary essential and non-essential amino acids on the nitrogen metabolism and showed that ammonia excretion depended on the ratio of essential to non-essential amino acids.

1.2.2 Pollution impacts on aquaculture

Aquaculture, especially mariculture is typically located in coastal areas. Through the intensified use and consequent pollution of coastal ecosystems by other stakeholders aquaculture production sites can be negatively influenced (Tisdell 1995; ICES Mariculture Committee 2003). Environmental risks originating from other users such as shipping, industrial and urban sewage influence the environment and therefore the water quality available for aquaculture production. Readman et al. (1993) focussed on the environmental distribution of tributyltin (TBT) a biocide which was added to marine paints as an antifoulant. They estimated that the use of TBT in Arcachon Bay (France) alone had led to a loss in revenue of 147 million U.S. dollars through reduced oyster production. Furthermore, Terlizzi et al. (1997) considered the morphological expression of imposex (the occurrence of penis and vas deference in females) in two species of muricids as a signal of a diffused TBT pollution along Italian coasts. Sawyer and Davis (1989) recovered and identified different species of terrestrial viruses, bacteria and protozoans from ocean waste disposals and sewage outfalls as these species represent excellent indicators for water and sediment contamination in marine ecosystems. Such impacts can be a direct or indirect thread to aquaculture species as they are exposed to chemicals, viruses, bacteria or other pollutants.

1.3 Mono-, Poly- and Integrated aquaculture

Aquaculture can be employed at different levels of production intensity. This can range from extensive production, relying on natural occurring food sources and applying low stocking densities, to intensive production with high stocking densities and supply of high energy food sources. Apart from the actual level of intensity, the number of cultured species in one production system can also vary.

1.3.1 Monoculture

Monoculture is defined as the production of one single species in an aquaculture system. Although, it is the most common system employed in conventional aquaculture production, the nutrient efficiency of such systems is considered to be low. The environmental impact of monoculture in open systems, such as net cages, can be substantial, especially to benthic organisms living on the sea or lake bed adjacent to cage facilities. Pearson and Rosenberg (1978) reported a gradual loss of benthic species as the degree of stress increased over space and/or time und cages. Because species differ in their tolerance to stress, there often is a pattern of replacement of the most sensitive species with more tolerant species as stresses begin and gradually increase. The abundance of the more tolerant species may initially increase as more sensitive species are excluded from the community, but they may eventually decline as the degree of stress continues to increase. Eventually, in highly polluted areas, no species will inhabit the sediments. The Pearson and Rosenberg model for benthic responses to stresses was based upon observations of organic enrichment of marine sediments. Numerous publications detailing with benthic responses to aquacultural pollution were published during the last decades (Enell and Loef 1983; Suvapepun 1994; Costa-Pierce 1996; Tovar et al. 2000; Jiang et al. 2004; Buschmann et al. 2006).

1.3.2 Polyculture

Polyculture, the production of several target species (e.g. fish, shrimps or crabs), that utilise different habitats and food sources in a single water body, provides an opportunity to improve the nutrient efficiency by internal recycling of nutrients within an aquaculture system. Species that feed on phyto- and zooplankton can be stocked with herbivorous and omnivorous species that feed at different levels of the food chain. Primary production from phytoplankton allows the recycling of excreted inorganic nutrients from animals inhabiting the same system and subsidises their own production. As a consequence, nutrient transfers within such a system can be balanced (Costa-Pierce 2002; McVey *et al.* 2002; Lucas and Southgate 2003; Lei 2006).

1.3.3 Integrated aquaculture

Integrated aquaculture represents a long-used form of culturing aquatic organisms. The concept of integrated aquaculture was historically used for the description of the co-culture of aquaculture and agriculture products (Kumar *et al.* 2000; Lucas and Southgate 2003; Andrew and Frank 2004). In this context integration represents the cultivation of various aquatic species in a single body of water, which is re-used for successive aquaculture species or even other crops, and combines aquaculture with other farm products or by-products (Lucas and Southgate 2003). The use of nutrient-rich effluents that originate from the production of terrestrial animals for fertilizing the water body and thus increasing the production of aquaculture is quite common.

In the context of the here presented work, integrated aquaculture will be referred to as the culture of aquatic organisms from different trophic levels in subsequent compartments of a recirculating aquaculture system which is operated totally independent from the natural environment. This concept is close to the idea of the conventional polyculture but it focuses more directly on culture of harvestable aquatic species from the wastewater stream of aquaculture without additional fertilisation and as a consequence reducing the concentrations of pollutants otherwise discharged to surrounding waters. Nutrient reduction is achieved by utilising dissolved and particulate nutrients for the production of autotrophic and detritivorous organisms. Such practises potentially include economical benefits for the operator as well as environmental benefits. With the same amount of nutrients a higher number of harvestable products can be achieved (Ryther 1983; Lin *et al.* 1993; Chopin *et al.*

2001; Davaraj 2001; Schneider et al. 2005). Integrated aquaculture can be applied for both open and closed systems but nutrient transfer as well as nutrient efficiency will be improved in closed recirculation systems. For conventional culture systems, such as ponds or net cages, the collection of solids is almost impossible, or extremely difficult. This is different in closed recirculating systems as the water extracted from the culture tanks can be fed through a sedimentation device to allow solids to settle and thereby be removed from the system's water. The next step can include a device for the culture of photoautotrophic organisms, such as algae, that assimilate and thereby remove dissolved nutrients from the system's water. Consequently, three harvestable products can be produced in one culture system, from a single application of feed to the key organism subsidised by the additional production of secondary organisms that utilise waste nutrients.

1.4 Outline of this work

At the start of this research gaps concerning the influence of detritivorous organisms on the performance of the recirculating system such as accumulation of dissolved and particulate nutrients and resulting oxygen depletion and H_2S-formation due to increased organic matter contents in the sediments were existing. Exact knowledge about the survival, growth and reproduction of detritivorous organisms under the applied conditions (e.g. amounts and quality of food) were also very limited. The performance of the sediment used simultaneously as sink for particulate nutrients and nitrification / dentrification unit was unknown.

Based on results and established experiences from former research, new experiments in land-based culture systems at different scales were designed, and performed and evaluated while focussing on the biology and ecology of detritivorous organisms. The marine polychaete *Nereis diversicolor* and the marine crustacean *Crangon crangon* were selected as suitable organisms for this research.

Experiences gained from small scale experiments were applied over a longer time period during the run of a newly designed Marine Recirculated Artificial Ecosystem (MARE) to test the performance of *N. diversicolor* as a secondary aquaculture product.

This work is divided into five chapters presenting the findings of this research. Additionally, a general introduction indicating the scientific knowledge at the start of the project and a combined conclusion of the findings from this research is also presented.

Major scientific objectives of this work were:

- To investigate the feasibility that the detritivorous polychaete *Nereis diversicolor* represents a suitable organism for the consumption and thereby reduction of solid wastes derived from recirculating aquaculture systems targeting primarily on the culture of carnivorous fish. For this purpose, growth and mortality were chosen as indicators for evaluating the feasibility of the use of the worms in aquaculture systems.

- To examine the effect of different diets on the fatty acid composition of the worms with possible implementations for aquaculture.

- To analyse the bioturbation effect caused by the polychaete within the culture tank sediments, which are derived from the waste of the carnivorous fish unit, while particularly focussing on the nitrification potential as well as the bacterial abundance and composition within the sediments inhabited by the worms.

- To test the applicability of a multitrophic integrated recirculating system designed for water and nutrient recycling and thereby optimizing water consumption of the recirculating system and simultaneously increasing the efficiency of nutrient uptake/recycling.

- To investigate the feasibility of the omnivorous crustacean *Crangon crangon* as an alternative culture organism for the consumption and thereby reduction of solid wastes derived from recirculating aquaculture systems besides the polychaete *N. diversicolor*.

Chapter 2

MARE – Marine Artificial Recirculated Ecosystem: implementation of a novel integrated recirculating system combining fish, worms and algae

Kube N., Bischoff A.A., Blümel M., Wecker B. and Waller U.

Abstract

Due to predicted limited resources in the future, such as clean water and nutrients, we investigated the implementation of a newly designed recirculating aquaculture system combining the requirements of closed recirculating systems for aquaculture with the demands of nutrient recycling within the "Marine Artificial Recirculated Ecosystem (= MARE)". The aim is, to support adequate fish growth as well as the production of secondary organisms and thereby maintain water quality within safe limits for all cultured organisms. In conventional recirculation systems, the majority of nutrients supplied with food are discharged because most of these installations do not include treatment steps for nutrient recycling. The MARE-system is based on the concept of a land based closed biological integrated seawater recirculating system allowing an extremely low water discharge (< 1 % d^{-1} system volume) managed by nutrient recycling units. Two consecutive long-term experiments were carried out investigating the performance of this advanced farming design with Gilthead sea bream (*Sparus aurata*, Sparidae) as the primary organism. In secondary cultivation compartments of the MARE-system, the detritivorous worm *Nereis diversicolor* (Polychaeta) was selected for the removal of particulate matter in both trials. Dissolved nutrients were utilised by the seaweed *Solieria chordalis* (Rhodophyta) during the first experimental trial (MARE I) or photobioreactors for the cultivation of the microalgae *Nannochloropsis* sp. (Chrysophyta) in combination with a conventional trickling biofilter during the second experimental trial (MARE II). Fish grew from 66 ± 13 g to 295 ± 42 g during MARE I and from 355 ± 49 g to 607 ± 91 g during MARE II. Maximum specific growth rates of the worms, calculated with the median, for subsequent experiments were 0.023 d^{-1} and 0.058 d^{-1}, respectively. Daily macroalgae yield was up to 171 g wet weight d^{-1} m^{-2} (μ_{max} = 0.025 d^{-1}). Maximum specific growth rates for the microalgae in continuous culture were 0.025 h^{-1}.

The general scientific concept turned out to be applicable to practice: integration of different trophic levels led to increased nutrient utilization. However, the volumes and stocking densities of all compartments have to be thoroughly adjusted to each other to optimize both primary and secondary production.

1. Introduction

In closed recirculation systems waste products are usually concentrated and discharged (solids) or are accumulating (dissolved nutrients such as nitrate and phosphate) in the system. The daily amount of produced waste is rather high: results from different aquacultural production systems showed that only a fraction of 20 – 30% of the nitrogen content of the feed is retained by fish (Krom *et al.* 1985; Krom and Neori 1989; Hall *et al.* 1992; Lupatsch and Kissil 1998; Hargreaves 1998). Utilization of phosphorus is in the range of 10 – 30% (Krom *et al.* 1985; Krom and Neori 1989; Barak and van Rijn 2000; Lupatsch and Kissil 1998). Thus, 70 – 90% of provided nitrogen and phosphorus are excreted either in dissolved or in particulate form.

To date, the operation of recirculation systems exclusively focuses on water recycling. Nutrients like nitrogen, phosphorus or carbon are eliminated by either biological water treatment steps (biofiltration) or by water exchange (Losordo *et al.* 1999; Waller *et al.* 2005). Consequently valuable organic compounds are lost from the system and the release can be regarded as an additional environmental burden (Ackefors and Enell 1994). In modern aquaculture systems a comprehensive nutrient recycling should be maintained in order to reduce environmental impacts. Chamberlain and Rosenthal (1995) argued, that waste from fish cultivation should be understood as a „new resource". Due to higher nutrient efficiencies achieved by the integration of different trophic levels, the profitability of recirculation systems can be enhanced (Asgard *et al.* 1999; Schneider *et al.* 2005). Chopin *et al.* (2001) found, that additional biomass supports the economical diversification and the benefit per production unit. Consequently, cultivation systems without any nutrient and energy recycling are supposed to have fewer chances in future (Troell *et al.* 2003).

The integration of secondary biological steps for the removal of nutrients and energy is not a new idea. In freshwater and brackish water systems integrated aquaculture has been practiced for centuries (Chopin *et al.* 2001). Marine seaweed production is the most prominent integration step applied in open marine culture systems (Petrell *et al.* 1996; Newkirk 1996; Chopin *et al.* 1999a; b) as well as in land-based aquaculture (Neori *et al.* 1991; 2000; Krom *et al.* 1995; Vandermeulen and Gordin 1990; Buschmann *et al.* 1996). Schneider *et al.* (2005) carried out a literature study concerning the knowledge and gaps of integrated aquaculture. Numerous organisms including aquatic plants and animals were investigated for secondary production steps in integrated aquaculture. However, all of these systems are characterized by comparably large water exchange rates. So far gaps concerning the knowledge about replacing conventional water treatment steps by secondary production units in recirculating aquaculture systems are still tremendous. The re-use of solid waste from recirculating systems has up to date hardly been applied for aquatic production. Consequently information about the nutrient recycling within land based marine recirculating integrated aquaculture systems is very limited.

The MARE experiments were conducted in order to fulfil three objectives of equal relevance: a) to develop the knowledge base required for the operation of an integrated marine recirculating system within safe limits for all cultured organisms, i.e. keeping concentrations of ammonia and nitrite low, maintaining pH values and oxygen saturation at constant levels and stabilizing the concentrations of nitrate and phosphate. Suitable secondary organisms were selected for the utilization of these nutrients. The other major objectives were b) to evaluate the feasibility of nutrient recycling within different components of the integrated recirculating system by simultaneously quantifying the growth of all cultured organisms and c) to provide an adequate data base from long term experiments for the development of a model describing the nutrient budget of the integrated aquaculture system.

2. Material and Methods

2.1 General description of the recirculation system

The MARE-system (= Marine Artificial Recirculated Ecosystem) represents the first attempt of an indoor low water discharge, multitrophic seawater recirculation system. The system consisted of several tanks and had a total volume of approx. 5 m^3 (Fig. 1). It consists of two self-cleaning conical fish tanks (1), one rectangular tank with baffle plates (2) and one circular tank (3), which were connected to form the recirculating system. A pump (4) (type AG8, ITT Hydroair international, Denmark) provided water to two foam fractionators (5) (type Outside Skimmer III; Erwin Sander Elektroapparate GmbH; Uetze-Eltze, Germany) as well as to the fish tanks. Tanks (2) and (3) received water from the fish tanks (1) by gravitational force. Table 1 presents the technical characteristics of all applied system compartments, such as surface area, water volume and adjusted flow rates.

The protein skimmers were supplied with compressed air and additional ozone, produced by an ozone generator (Sander, Ozonizer A2000). A secondary freshwater loop, not included in Fig. 1, containing a holding tank and a pump, was attached to the foam fractionators. The function of this secondary water loop was the automatical rinsing and collection of foam produced by the foam fractionators and thereby collecting the solids removed from the system water.

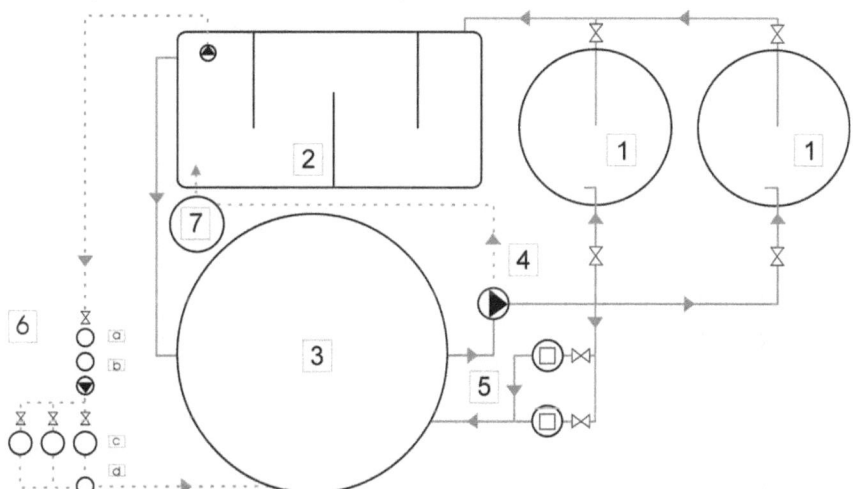

Fig. 1: Flow-chart of the MARE-system. Solid lines = configuration of the first experimental trial (MARE I), dashed lines = additional components during the second experimental trial (MARE II). (1) = self-cleaning tanks used for the cultivation of fish during both experiments, (2) = tank used for the cultivation of *N. diversicolor* in both trials; (3) = tank used for the cultivation of the seaweed *S. chordalis* during MARE I and for the cultivation of fish during MARE II; (4) = pump; (5) = foam fractionators; (6) = photobioreactor system for the cultivation of *Nannochloropsis* sp. during MARE II: a) pre-treatment unit, b) degassing tower, c) culture units, d) harvesting unit; (7) = trickling biofilter used during MARE II. Double triangle = tap, arrows = water flow. Dimensions of the components are given in the Tab. 1.

Tab.1: Description of the different compartments of the MARE-system. Water volume and surface area of each compartment are presented for individual compartments such as fish tanks, foam fractionator, microalgae reactor.

system compartment	quantity [n]	water volume [L]	surface area [m²]	adjusted flow rates [L h⁻¹]
fish tank	2	700	0.79	700
worm tank	1	1350	2.08	1400
macroalgae tank	1	2000	2.30	3400
microalgae reactor	3	50	0.03	1.5
foam fractionator	2	55	0.03	1000
biofilter	1	30	35	500 - 1000

Two long-term experiments with different modifications of the system components (details described in 2.1.1 and 2.1.2) were performed: the first trial was done from November, 11th, 2004 until June, 16th, 2005 (217 experimental days) and will be referred to as MARE I. The second trial is indicated as MARE II and was realized from September, 5th, 2005 until February, 15th, 2006 (163 experimental days).

2.1.1 System configuration of MARE I

During MARE I the recirculation system (Fig. 1, solid lines) consisted of two self-cleaning conical tanks (1). Each tank was stocked with 85 juvenile Gilthead sea bream (*S. aurata*, Sparidae) with initial fish weights in the range of 30 to 99 g; average and standard deviation amounted to 66 ± 13 g. The resulting initial stocking density was 2.5 kg m^{-3} system volume. Tank 2 was filled with a 0.1 m deep sand layer (grain size ≤ 2 mm; water column 0.7 m) and stocked with the detritivorous worm *N. diversicolor* (Polychaeta) at a stocking density of 900 – 950 individuals per m². Tank 3 was used for the cultivation of macroalgae. This tank was provided with a central aeration device and artificial illumination of 400 µE m^{-2} s^{-1} with a day/night cycle of 16:8 hours. The seaweed *S. chordalis* (Rhodophyta) was cultivated free floating with an initial biomass of 7.3 kg m^{-2} tank surface area.

2.1.2 System configuration of MARE II

Fish biomass was increased compared to MARE I; the conical tanks (1) were each stocked with 35 Gilthead seabream (*S. aurata*) and 65 additional animals were cultivated in tank 3. Initial fish weights were in the range of 205 to 460 g; average weight and standard deviation amounted to 355 ± 49 g. The resulting initial stocking density was 9.9 kg m^{-3} system volume. Tank 2 was additionally stocked with a second generation of *N. diversicolor*, resulting in an estimated initial abundance of approx. 850 individuals per m².

During MARE II two additional components were included (Fig. 1, dashed lines): A photobioreactor system (6a – d) for the continuous cultivation of the microalgae *Nannochloropsis* sp. (Chrysophyta) was integrated as a bypass between tanks 2 and 3. The photobioreactor system consisted of a disinfection unit, a cultivation and a harvesting unit. In the disinfection unit, water from the recirculation system was pre-treated in a foam fractionator (6a) at high redox potential values. After this pre-treatment, the water was transferred to a degassing tower (6b) with aeration to remove residual ozone. *Nannochloropsis* sp. was cultivated in three acrylic columns (6c). The columns were equipped with light tubes (100 – 400 µE m^{-2} s^{-1}) and air diffusers. The water flow rate through the photobioreactors was adjusted according to the growth rate of the microalgae. The microalgae were continuously harvested (6d) and treated water flowed back into the main system.

For further details concerning the design and performance of the photobioreactor system, we refer to Kube (2006). Additionally a trickling biofilter (7) was installed in order to maintain low ammonia and nitrite concentrations.

2.2 Measurements and Methods

2.2.1 Chemical parameters of the water

Water samples for analysis of dissolved nutrients (PO_4-P, NO_3-N, NO_2-N and Total Ammonia Nitrogen = TAN) were taken daily from the outlets of the fishtanks, the worm tank and the macroalgae tank. During MARE II the outlet of the biofilter was also sampled. Water samples were stored at –20 °C for one month and analysed colorimetrically by a continuous flow analyser (AA3 Bran+Luebbe, Norderstedt, Germany).

Online measurements of redox potential, pH and dissolved oxygen were recorded with a control module KM 2000 (Sensortechnik Meinsberg GmbH, Meinsberg, Germany) and a portable measuring device (WTW multi 350, Weilheim, Germany). Salinity of the system was adjusted around 25.

2.2.2 Solid components

Samples from the rinsing water of the foam fractionators (secondary freshwater loop) were taken weekly to determine the amount of removed solids. 10 ml samples (n = 12) were centrifuged and the supernatant was stored for later analysis of dissolved nutrients. Sub-samples (n = 12) of the microalgae harvest were centrifuged (5000 rpm, 10 minutes).

Furthermore, the total organic matter (TOM) of the sediment was analysed on a weekly base using sub-samples of approx. 10 cm³ sediment (n = 3 for the first experimental trial and n = 5 for the second experimental trial) from different sampling points of the worm tank. Therefore the tank was divided into either 3 sections (MARE I) or 5 sections (MARE II) and from each section one sediment sample was collected. Sampling point 1 was always located in the first section of the tank where the water inlet was located (Fig. 2).

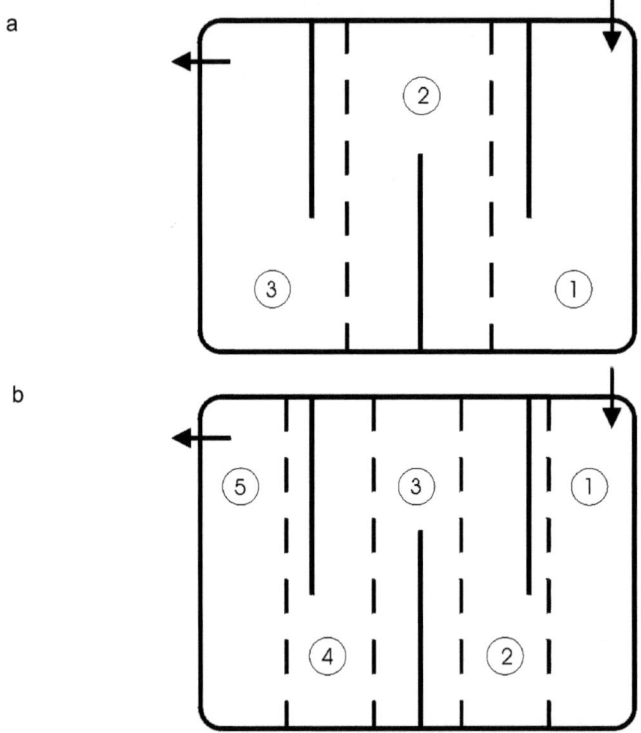

Fig. 2: Sampling areas for the analyses of total organic matter in the sediment of the *Nereis* bioreactor. The inward arrows indicate the location of the water inlet. Samples were collected randomly from the entire area. Sampling point 1 was always located in the section with the water inlet. During MARE I (a) the tank area was divided into three sections and during MARE II (b) it was divided into five sections.

Dry matter content, TOM and energy were measured according to Winberg and Duncan (1971). Dry matter content of all collected samples was determined after dehydration at 60 ± 5 °C overnight. TOM was analysed by incineration at 450 ± 50 °C in a muffle furnace for 24 hours. Energy content was measured by complete sample combustion using an IKA calorimeter C4000 (IKA, Staufen, Germany). C/N ratios were determined by gas chromatography (GC) in an element analyser (EURO EA elemental analyser, Milano, Italy).

2.2.3 Biomass determination

At intervals of approx. 3 weeks, a sub-sample of fish was measured to determine the fish biomass. Therefore, fish were sedated with clove oil and wet weight of individual fish was recorded. Pelleted fish feed (Biomar *Aqualife* 17) was supplied according to feeding tables. These tables apply the average fish size and water temperature to determine the feeding rate and by applying the actual fish biomass and the feeding rate the absolute amounts of fish feed can be calculated. The food conversion ratio was calculated by using the equation:

$$FCR = feed\ intake\ (g) / body\ weight\ gain\ (g) \qquad (1)$$

Worm biomass was determined (number and wet weight of worms) by using sediment sub-samples (n = 4) from the worm tank. Sediment cores of approx. 800 cm^3 were sampled. Sediment including the worms was sucked up with a hose and transferred onto a sieve (mesh size approx. 1 mm). Sediment was washed through the sieve with additional system water and worms remaining on the sieve could be collected. Specific growth rate µ of the worms was calculated by using the equation:

$$\mu = ln\ (W_t / W_0)^* t^{-1} \qquad (2)$$

where W_0 and W_t are average body mass (wet weight) of the polychaetes on Day 0 and Day t respectively.

Wet weight of *S. chordalis* was determined every 1 to 2 weeks. Biomass yield of *Solieria* remained in the tank until a stocking density of 8 kg m^{-2} was reached. This stocking density presents the optimum density for *Solieria* (Sylter Algenfarm pers. comm.), below this density epiphytes will develop and beyond this density self induced light limitation for the macroalgae will occur. After reaching the optimum stocking density, the macroalgae biomass was kept constant at this value by removing the weekly gain from the system.

The microalgal biomass in the photobioreactors was determined using optical density measurements (HACH SR2010 photometer, wavelength 665 nm). For that purpose a calibration curve was developed by counting cell abundances and associated optical density measurements. Daily yield of microalgae was determined by using the harvested volume and the measured optical density of the harvest.

3. Results

3.1 Chemical parameters of the water

Water replacement rate throughout both experiments was less than 1% per day (system volume).

Fig. 3 shows the concentrations of dissolved nutrients during MARE I (with macroalgae filter).

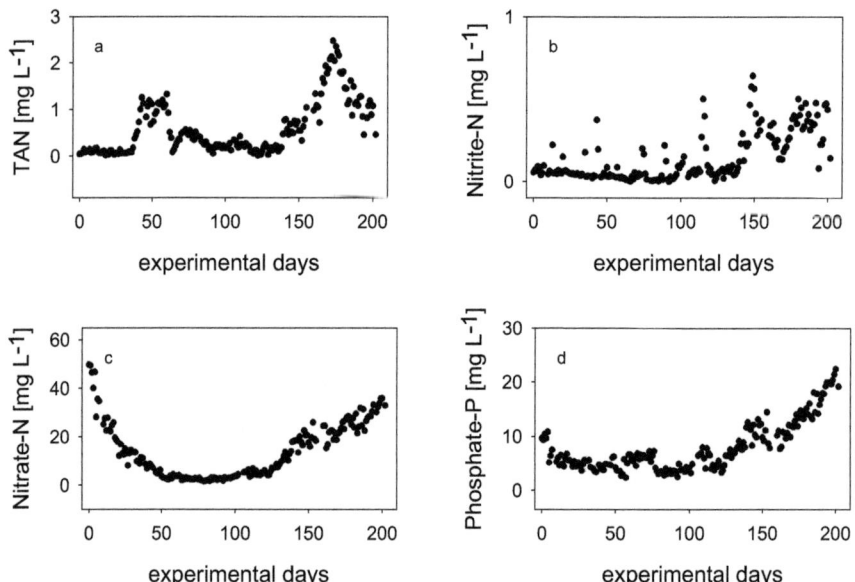

Fig. 3: Dissolved nutrients obtained during MARE I: a) TAN; b) Nitrite-N; c) Nitrate-N and d) Phosphate-P.

Until experimental day 150, TAN concentrations were most of the time low with a peak of 1.32 mg L^{-1} between experimental day 45 and 63 (Fig. 3a). At experimental day 139 of MARE I the TAN concentrations started to rise up to 2.47 mg L^{-1} and decreased from day 174 onwards to levels around 0.50 mg L^{-1} at the end of the experiment.

Concentrations of nitrite-N were at constant low levels during most of the experiment ranging from 0 to 0.21 mg L^{-1} with few peaks above average (Fig. 3b). Increasing nitrite-N concentrations could be detected after day 145, reaching values around 0.44 mg L^{-1} and decreasing concentrations at the end of the first experiment.

During MARE I concentrations of nitrate-N were decreasing from 49.74 mg L^{-1} (experimental day 1) below 5 mg L^{-1} until experimental day 122 (Fig. 3c). From there on concentrations were continuously rising up to values of 35 mg L^{-1} until experimental day 200. Similar values could be observed for the concentrations of orthophosphate-P with values below 8 mg L^{-1} until experimental day 136, increasing afterwards to concentrations up to 22 mg L^{-1} (Fig. 3d).

During MARE II concentrations of dissolved nutrients were higher compared to MARE I due to the increased fish biomass in the system (Fig. 4).

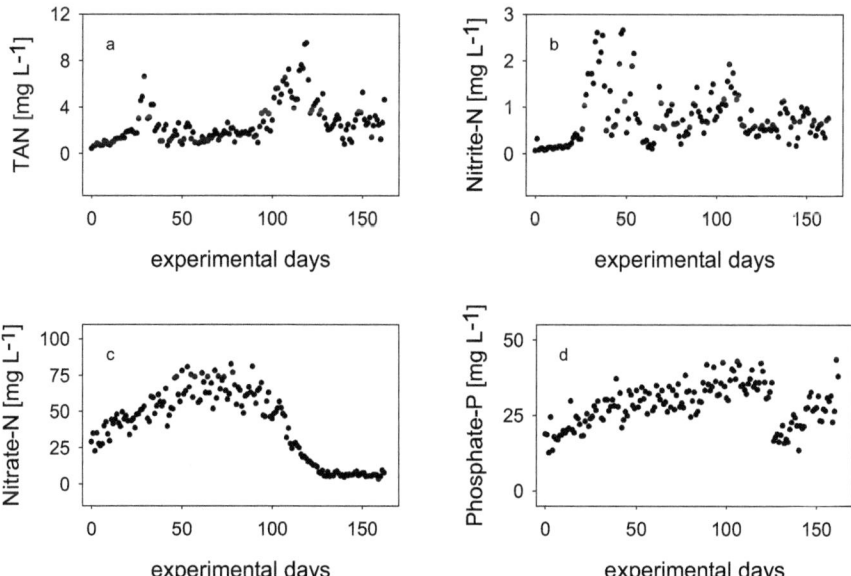

Fig. 4: Dissolved nutrients obtained during MARE II: a) TAN; b) Nitrite-N; c) Nitrate-N and d) Phosphate-P.

TAN concentrations showed a constant increase starting at 0.42 mg L^{-1} and reaching values around 2.64 mg L^{-1} at the end of the second experiment. Two distinct TAN peaks around experimental day 29 (6.62 mg L^{-1}) and 119 (9.51 mg L^{-1}) were recorded (Fig. 4a). Nitrite-N concentrations were slightly increasing throughout the entire experimental period with some variations (Fig. 4b). Starting values were 0.06 mg L^{-1} and final values reached up to 0.76 mg L^{-1}, with peaks at experimental days 34, 37 and 48 showing peak values of 2.60, 2.54 and 2.66 mg L^{-1}, respectively. Nitrate-N and orthophosphate-P were accumulating in the system reaching maximum values of 82.80 mg L^{-1} for nitrate-N and 43.50 mg L^{-1} for orthophosphate-P, respectively. Due to the prolonged second peak of TAN a water exchange of 1000 L was done, reflected by the steep decrease of phosphate concentrations at day 120 (Fig. 4b – d). From experimental day 90 onwards a decrease of nitrate-N could be observed, resulting in low values around 5 mg L^{-1} at day 162.

No fish mortality caused by elevated TAN or nitrite concentrations were observed during both experimental trials.

During MARE I pH values were 7.97 ± 0.24 but showed a slight decrease at the end of the experimental period (Fig. 5a). Throughout the first period of MARE II pH values decreased from 7.5 to 6.1 and an increase of pH by the addition of CaO was necessary. After experimental day 110 the pH had to be stabilised with HCl (Fig. 5b).

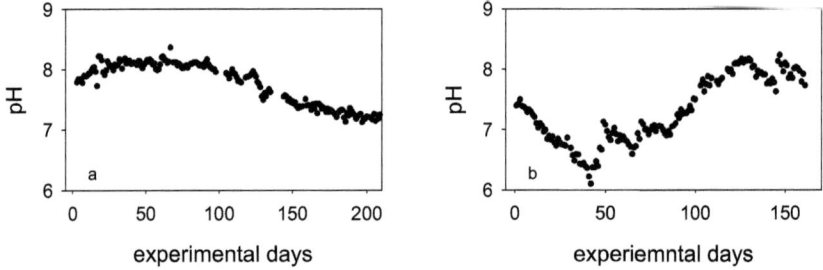

Fig. 5: pH values obtained during both MARE experiments: a) pH during MARE I; b) pH during MARE II.

Abiotic parameters for both experimental trials are shown in Tab. 2. Water temperature, oxygen saturation and salinity did not differ between both experimental periods.

Tab. 2: Abiotic parameters (means ± SD) - water temperature [°C], oxygen saturation [%] and salinity – of both long term experiments performed in the MARE system [n ≥ 30].

	water temperature [°C]	oxygen saturation [%]	salinity
MARE I	19.9 ± 0.5	74.0 ± 15.5	26.3 ± 1.4
MARE II	19.3 ± 0.5	72.7 ± 8.6	24.8 ± 1.0

3.2 Growth performance

Tab. 3 reflects the feeding days and the daily amounts of food supplied to the fish during both experiments.

Tab. 3: The feeding periods of the two MARE experiments, the corresponding feeding days and the daily amounts of food supplied to the fish.

	MARE I			MARE II	
period	feeding days	daily amount of food [g]	period	feeding days	daily amount of food [g]
11.11.04 – 01.12.04	17	58	05.09.05 – 23.09.05	18	222
02.12.04 – 22.12.04	19	73	24.09.05 – 13.10.05	19	468
23.12.04 – 14.01.05	21	84	14.10.05 – 03.11.05	26	521
15.01.05 – 11.02.05	27	88	04.11.05 – 06.12.05	28	514
12.02.05 – 14.03.05	26	112	07.12.05 – 09.01.06	34	503
15.03.05 – 13.04.05	29	135	10.01.06 – 27.01.06	18	570
14.04.05 – 24.05.05	40	162	28.01.06 – 15.02.06	19	618
25.05.05 – 15.06.05	20	206			

The weight data obtained from the measurements of *S. aurata* in the MARE-system during MARE I are given in Tab. 4.

Tab. 4: Average fish weight [g] and stocking densities [kg m^{-3} system volume] of *S. aurata* cultured in the MARE system during the first experimental phase (MARE I).

date	experimental day	n	average weight ± SD [g]	stocking density [kg m^{-3} system volume]
11.11.2004	0	40	66 ± 13	2.5
01.12.2004	20	114	74 ± 12	2.8
22.12.2004	41	61	86 ± 12	3.4
14.01.2004	64	60	105 ± 15	4.0
11.02.2005	92	51	132 ± 16	5.0
14.03.2005	123	50	161 ± 21	6.1
13.04.2005	153	51	198 ± 29	7.5
24.05.2005	194	63	252 ± 33	9.6
16.06.2005	217	51	295 ± 42	11.1

Within 217 days fish grew on average from 66 ± 13 g to 295 ± 42 g (mean weight ± standard deviation), resulting in a maximum stocking density of 11.1 kg m^{-3} (system volume). Average food conversion ratio (FCR, ± SD) during MARE I was 1.09 ± 0.32 and ranged from 0.71 to 1.64.

In MARE II (Tab. 5) fish nearly doubled their mean weight (± standard deviation) from 355 ± 49 g to 607 ± 91 g at maximum stocking densities of 16.4 kg m^{-3} (system volume). FCR during MARE II varied between 1.09 and 5.46.

Tab. 5: Average fish weight [g] and stocking densities [kg m^{-3} system volume] of *S. aurata* cultured in the MARE system during the second experimental phase (MARE II).

date	experimental day	n	average weight ± SD [g]	stocking density [kg m^{-3} system volume]
05.09.2005	0	127	355 ± 49	9.9
23.09.2005	18	57	375 ± 39	10.4
13.10.2005	38	45	417 ± 47	11.6
03.11.2005	59	68	458 ± 82	9.7
06.12.2005	92	57	485 ± 69	12.5
09.01.2006	126	59	552 ± 83	14.1
27.01.2006	144	53	594 ± 76	15.3
15.02.2006	163	63	607 ± 91	16.4

Tank 2, the *Nereis* bioreactor, was stocked with a total of approx. 1900 individuals of *N. diversicolor* in June 2004. Total initial worm biomass was 1.8 ± 0.5 kg. Worm weight (median; P10 and P90 values) was 887.3 mg (557.0; 1242.2) (n = 270). Abundance determination at the start of the first MARE experiment (day 0, Fig. 6) revealed worm numbers exceeding 40.000 individuals per m^2 in the bioreactor (Fig. 7), representing a total biomass of 4.4 ± 3.7 kg (mean ± SD). Individual worm weight (median; P10 and P90 values) was 68.2 mg (25.6; 159.1) (n = 150) at the start of the experiment, clearly indicating a reproduction event between the biomass determinations.

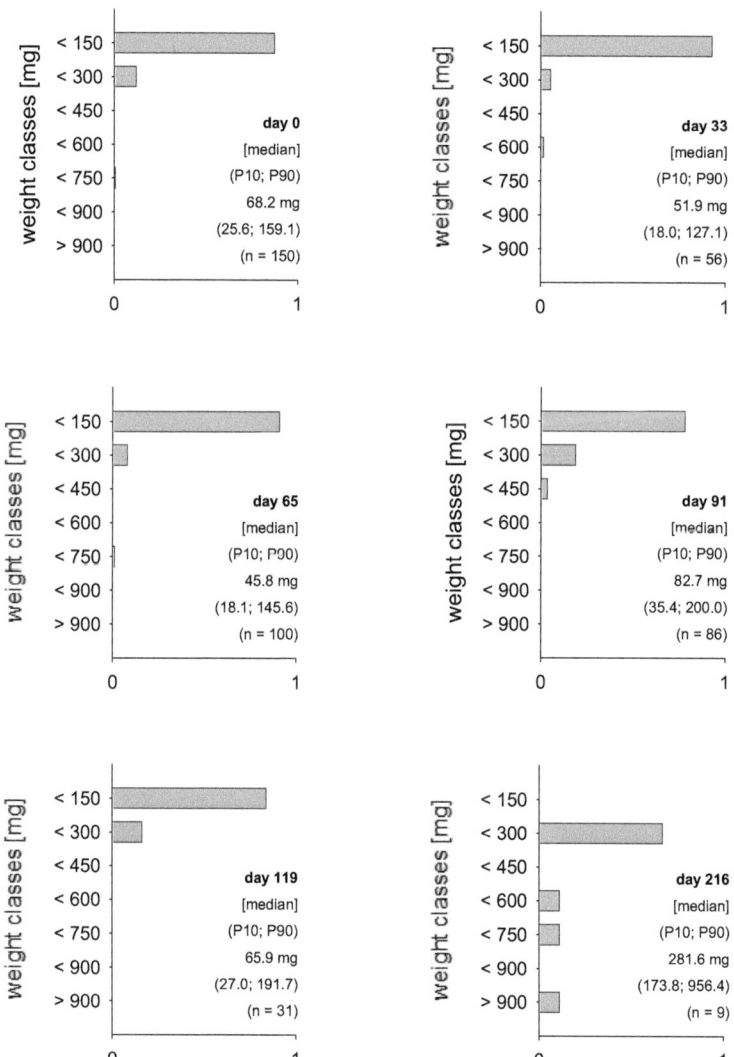

Fig. 6: Weight class distribution of collected and weighed *N. diversicolor* during MARE I. The experimental day and the wet weight (median; P10 and P90 values) in [mg] of each sampling are also presented in each graph.

Worm abundance and total biomass (abundance x average weight of the worms) in the bioreactor showed a continuous decrease during the experimental period (Figs. 6

and 7). The average worm weight was constantly low until biomass determination at experimental day 119 and increased at the end of the experiment. Maximum specific growth rates of 0.023 d^{-1} could be observed.

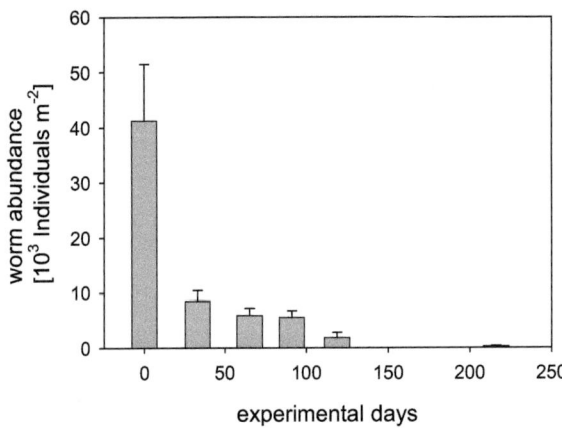

Fig. 7: Worm abundance (mean ± SD) recorded during sampling of MARE I. (n = 4)

Total organic matter (TOM) of the sediment samples during MARE I showed low values compared to natural conditions. Besides few exceptions, TOM was around 2% (Fig. 8).

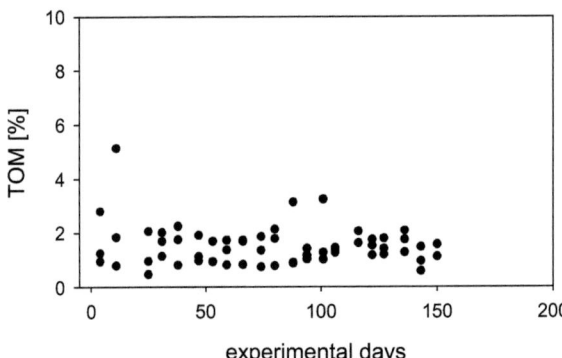

Fig. 8: Total organic matter content of the sediment obtained from the MARE system during the first experimental period (MARE I).

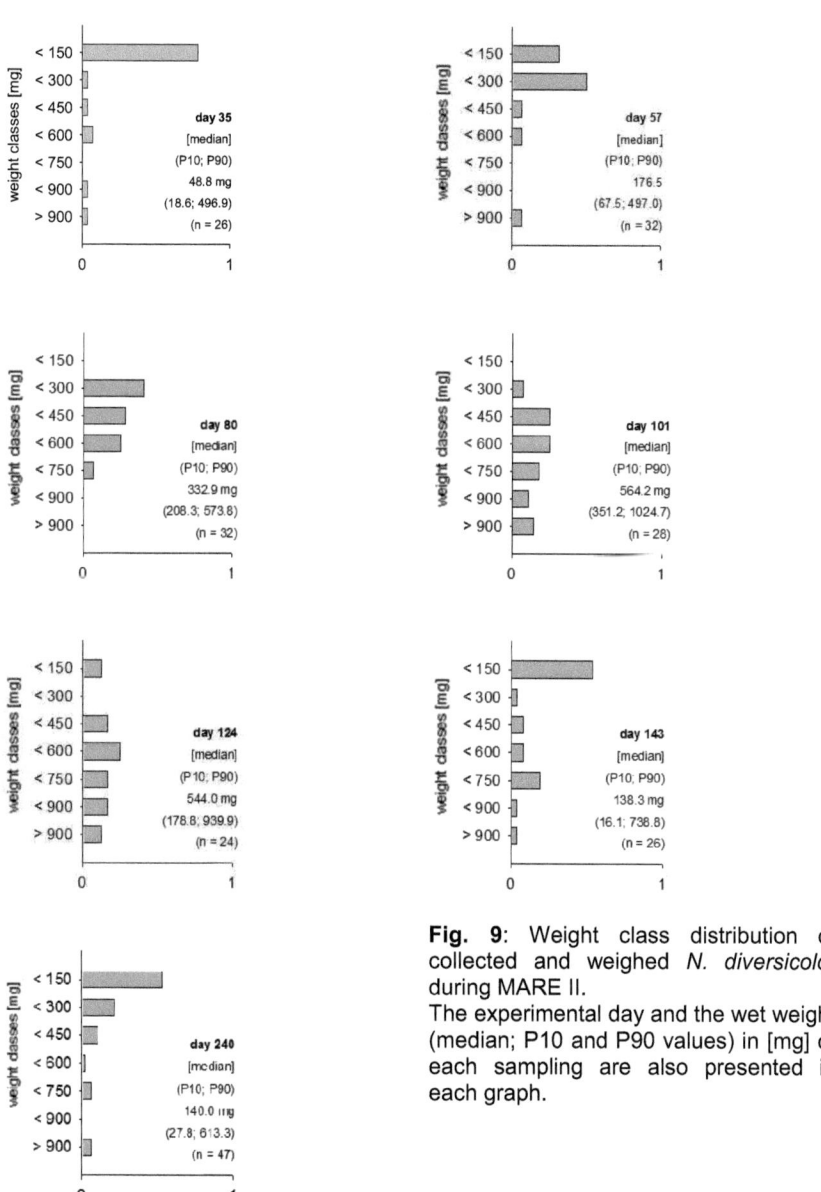

Fig. 9: Weight class distribution of collected and weighed N. diversicolor during MARE II.
The experimental day and the wet weight (median; P10 and P90 values) in [mg] of each sampling are also presented in each graph.

During MARE II individual worm weight constantly increased from 48.8 mg (18.6; 496.9) (median, P10 and P90 values) to maximum values of 564.2 mg (351.2; 1024.7) at experimental day 101 and 544.0 mg (178.8; 939.9) at day 124, respectively (Figs. 9 and 10). Maximum specific growth rates of 0.058 d^{-1} could be achieved. After experimental day 100 a reproduction event occurred within the worm tank. Fig. 9 shows a decrease of average worm weight from day 101 onwards. The graph also presents the weight class distributions of the sampled worms. After day 143 the majority of worms were measured in the smallest weight class and the fraction of larger individuals decreased during the remaining experimental phase. Worm abundance during MARE II was between 850 and 1600 Individuals per m² (Fig. 10).

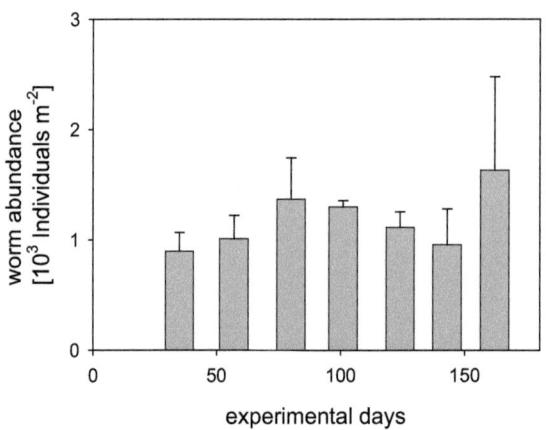

Fig. 10: Worm abundance (mean ± SD) recorded during sampling of MARE II. (n = 4)

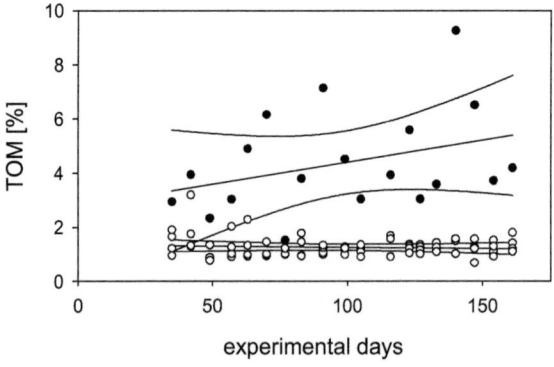

Fig. 11: Total organic matter content of the sediment obtained from the MARE system during the second experimental period (MARE II). (● = sampling point 1; ○ = sampling points 2 – 5)

TOM of the sediment during MARE II (Fig. 11) showed significant differences between sampling point 1, close to the water inlet, and the other sampling points in the bioreactor of the worms (ANOVA, $p < 0.001$). Total organic matter of the first sampling point constantly increased during the experiment, whereas the other sampling points showed low concentrations during most of the experimental period. The growth performance of *S. chordalis* within the MARE system is shown in Tab. 6 (Part I and II). At experimental day 1 (11.11.2004) *Solieria* was initially stocked. Recommended stocking density was 8 kg FW m^{-2} surface area (Sylter Algenfarm, pers. comm.). For the algae tank of the MARE-system a total biomass of 18 kg FW was recommended. Due to a shortage in the supply of stocking material a reduced stock of 16.8 kg FW was used for the experimental trial and due to reduced growth of *Solieria*, recommended stocking density could not be achieved during the following weeks of the experiment. On experimental day 60 (10.01.2005) the macroalgae tank was restocked with new algae at reduced stocking densities. Growth performance of *Solieria* during the second stage of MARE I improved and from mid February onwards (18.02.2005, experimental day 99) recommended stocking density of 18 kg could be realised. Macroalgae yields between 28 and 171 g d^{-1} m^{-2} could be achieved during the second stage of the experiment and resulted in a maximum specific growth rate of 0.025 d^{-1}.

Tab. 6 – Part I: Initial and final biomass of the first stocking of *S. chordalis* during MARE I. Biomass figures represent fresh weight.

experimental period	days	initial biomass [g]	final biomass [g]	growth rate [g FW d^{-1} m^{-2}]	specific growth rate µ [d^{-1}]
10.11.04 – 17.11.04	6	16800	17000	15	0.0020
17.11.04 – 24.11.04	6	15600	15731	10	0.0014
24.11.04 – 10.01.05	46	14331	15000	6	0.0010

Tab. 6 – Part II: Initial and final biomass of the second stocking of *S. chordalis* during MARE I. Biomass figures represent fresh weight.

experimental period	days	initial biomass [g]	final biomass [g]	growth rate [g FW d^{-1} m^{-2}]	specific growth rate µ [d^{-1}]
10.01.05 – 21.01.05	10	11195	11836	28	0.0056
21.01.05 – 02.02.05	11	11836	15649	151	0.0254
02.02.05 – 09.02.05	6	15649	17201	113	0.0158
09.02.05 – 18.02.05	8	17201	20069	156	0.0193
18.02.05 – 25.02.05	6	18000	18453	33	0.0041
25.02.05 – 03.03.05	5	18000	18785	68	0.0085
03.03.05 – 11.03.05	7	18000	20758	171	0.0204
11.03.05 – 17.03.05	5	18000	18682	59	0.0074
17.03.05 – 30.03.05	12	18000	21054	111	0.0131

The daily harvest of microalgae averaged (± SD) 2.49 ± 1.34 g DW (n = 23) (Tab. 7 a-c)). The harvesting process was very efficient: daily harvest volume was 0.184 ± 0.053 L with high cell densities of 2.28 x 10^{10} ± 6.7 x 10^{8} cells ml^{-1}. 99.68 ± 0.38% of the water passing the photobioreactors was re-transferred to the main system. The energy content of the microalgae were 21.09 ± 4.6 KJ g^{-1} DW (n = 35) and the total amount of fatty acids varied from 5.5 ng fatty acids FA/µg POC to 16.72 ng FA/µg POC according to the varying irradiance supply for each photobioreactor. Fractions of arachidonic acid (AA) and eicosapentaenoic acid (EPA) ranged from 4.0 – 8.27% and 9.1 – 18.9% of the total fatty acid composition, respectively.

Nutrient uptake rates have been determined during 1 hour growth experiments revealing an average uptake of 0.19 – 0.29 mg h^{-1} and 0.13 – 0.19 mg h^{-1} for PO_4-P and NO_3-N per litre culture volume according to the applied light intensities (see Kube 2006).

Tab. 7a: Flow rates during the experimental period of the photobioreactor system as well as optical density measurements (OD_{665}) of the outflow from the harvesting unit.

Date	Flow rate per day [L]	OD_{665} (after harvest)
14.10.05	108.0	-
15.10.05	108.0	0.062
19.10.05	108.0	0.138
20.10.05	108.0	0.026
27.10.05	73.4	0.136
28.10.05	73.4	0.102
29.10.05	73.4	0.020
31.10.05	73.4	0.172
01.11.05	73.4	0.002
02.11.05	73.4	0.000
18.11.05	108.0	0.001
23.11.05	86.4	0.001
24.11.05	86.4	0.000
25.11.05	86.4	0.003
27.11.05	86.4	0.002
28.11.05	86.4	0.000
18.01.06	60.5	0.009
19.01.06	14.4	0.013
20.01.06	21.6	0.013
21.01.06	21.6	0.011
22.01.06	21.6	0.012
23.01.06	21.6	0.017
24.01.06	21.6	0.008
04.02.06	14.4	0.025
Average		
SD		

Tab. 7b: Differences of the inflow and outflow concentrations of the photobioreactor system in accordance to the data presented in Table 7a.

Date	Δ inflow – outflow conc. [mg L^{-1}]			
	PO$_4$-P	NO$_3$-N	TAN	NO$_2$-N
14.10.05	2.1	3.8	-	-
15.10.05	3.6	5.1	-	-
19.10.05	8.2	18.0	1.14	0.00
20.10.05	4.8	12.5	0.58	0.16
27.10.05	14.5	20.4	1.04	0.02
28.10.05	11.0	26.0	1.38	0.17
29.10.05	14.3	19.5	1.7	0.08
31.10.05	13.8	13.6	0.68	0.08
01.11.05	14.1	5.3	0.1	0.3
02.11.05	15.7	14.0	0.2	0.2
18.11.05	13.9	2.5	0.0	0.1
23.11.05	5.6	1.6	0.1	0.04
24.11.05	14.6	2.8	0.0	0.03
25.11.05	19.7	9.6	0.7	0.01
27.11.05	25.2	2.3	0.0	0.0
28.11.05	25.4	2.2	0.0	0.0
18.01.06	28.1	0.6	1.31	0.02
19.01.06	23.6	2.9	3.19	0.00
20.01.06	23.9	2.9	3.2	0.00
21.01.06	17.6	3.7	2.13	0.06
22.01.06	11.3	3.8	1.94	0.03
23.01.06	12.0	1.7	1.62	0.00
24.01.06	11.0	3.5	0.22	0.07
04.02.06	3.9	0.6	3.8	0.02
Average				
SD				

Tab. 7c: Daily amount of removed nutrients of the photobioreactor system in accordance to the data presented in Table 7a and 7b.

Date	amount of removed nutrients [g d^{-1}]			
	PO$_4$-P	NO$_3$-N	TAN	NO$_2$-N
14.10.05	0.23	0.41		
15.10.05	0.39	0.55		
19.10.05	0.89	1.94	0.12	0.00
20.10.05	0.52	1.35	0.06	0.02
27.10.05	1.06	1.50	0.08	0.00
28.10.05	0.81	1.9	0.1	0.01
29.10.05	1.05	1.43	0.13	0.01
31.10.05	1.01	1.0	0.05	0.01
01.11.05	1.03	0.39	0.01	0.02
02.11.05	1.15	1.03	0.02	0.02
18.11.05	1.5	0.27	0.00	0.01
23.11.05	0.48	0.14	0.01	0.00
24.11.05	1.26	0.24	0.00	0.00
25.11.05	1.7	0.83	0.06	0.00
27.11.05	2.19	0.20	0.00	0.00
28.11.05	2.19	0.19	0.00	0.00
18.01.06	1.7	0.04	0.08	0.00
19.01.06	0.34	0.04	0.05	0.00
20.01.06	0.52	0.06	0.07	0.00
21.01.06	0.38	0.08	0.05	0.00
22.01.06	0.24	0.08	0.04	0.00
23.01.06	0.26	0.04	0.04	0.00
24.01.06	0.24	0.08	0.01	0.00
04.02.06	0.06	0.01	0.06	0.00
Average	**0.88**	**0.57**	**0.05**	**0.00**
SD	*0.63*	*0.63*	*0.04*	*0.01*

4. Discussion

The MARE experiments were conducted a) to develop the knowledge base required for the operation of an integrated marine system within safe limits for all cultured organisms; b) to evaluate the feasibility of nutrient recycling within different components of the integrated recirculating system and c) to provide an adequate date base from long term experiments for the development of a model describing the nutrient budget of the integrated aquaculture system. The results are discussed here in detail with regard to aims a) and b). Aim c) is the focused subject of another publication by Wecker (2006.).

Performance of the MARE-system

The primary aim of this study was to improve the knowledge required for a safe operation of the MARE-system. Maximum stocking densities of 11.1 kg m^{-3} (system volume) (MARE I) and 16.4 kg m^{-3} (system volume) (MARE II) were recorded and these values are according to Colloca and Cerasi (2005) in the lower range of stocking densities for Gilthead sea bream reared in conventional aquaculture tanks which range from 15 – 45 kg m^{-3}.

Although the concentrations of inorganic dissolved nutrients (ammonia and nitrite) as well as oxygen levels would have allowed a biomass increase in the fish tanks, the volume of these tanks limited the number of fish during MARE I. In order to analyse the system stability with respect to increased fish biomass, during MARE II tank 3, which was used for the cultivation of macroalgae during MARE I, was also stocked with fish. Due to the stocking of tank 3, suspended solids from the fish were directly transferred to the foam fractionators. At the end of the second experimental period the functionality of foam fractionation was restricted due to large quantities of suspended solids caused by the missing first step of solid removal.

Nevertheless, fish growth achieved in the system was sufficient and animals reached commercial size of approx. 350 g per individual after 14 – 15 months. The reduced growth compared to other systems (12 – 13 months) (Lupatsch and Kissil 1998; Lupatsch et al. 2003; Colloca and Cerasi 2005) was assumed to be caused by different reasons. These reasons included lower water temperatures and consequently lower feeding rates, reduced feeding caused by biomass determinations as well as maintenance work during the experiments. During MARE II fish growth was decreasing, reaching maximum specific growth rates of 0.0056 d^{-1} compared to MARE I with maximum specific growth rates of 0.0179 d^{-1}. This can be explained by the increased body weight of the fish. Average body weight of the fish at the end of MARE II was almost twice as high as the average body weight at the end of MARE I.

Noticeably, feed conversion ratio (FCR) during the complete experimental period of MARE I was very good and in the range described for commercial farms (Colloca and Cerasi 2005). The average FCR value of 1.1 is below values described by Lupatsch and Kissil (1998) (FCR = 1.5) and Porter et al. (1986) (FCR = 2.7). This might be due to enhanced efforts of fish feed producers in the last few years to adapt feed compositions better to the requirements of the cultured species. During the experimental period of MARE II under regular conditions FCR increased slightly compared to MARE I. These increases in FCR were comparable to commercial farms and can be explained by the physiology of fish, increasing FCR with increasing body weight which was also reported by Imsland et al. (2006) who investigated the effects of fish size on the feed efficiency. During two periods of MARE II the FCR showed high values and these values were connected to elevated TAN concentrations. This may have led to behavioural changes, caused by unionised ammonia which is toxic at higher concentrations, and consequently led to reduced feed intake by the fish.

To ensure system stability, the control of fish faeces played a key role due to its crucial impact on water quality (Cripps and Bergheim 2000). The MARE-system included a two step solid separation (sedimentation and foam fractionation) to ensure best water quality (Losordo et al. 1999; Summerfelt 2002). In contrast to conventional systems particulate waste (fish faeces and remaining feed material) was not eliminated from the integrated system and as a result leaching occurred within the first 6 hours (Lupatsch and Kissil 1998) and could hardly be influenced. However, critical concentrations of dissolved nutrients or oxygen saturation could not be observed within the water column of the MARE-system at any time of the experimental periods.

The decrease of pH at the end of the experimental period (MARE I) can be explained by the lower growth performance of the macroalgae. Consequently, the uptake of CO_2 by the macroalgae was not sufficient to stabilise pH values. Thus, CO_2 reacts in the water column and, depending on the pH of the water, forms bicarbonate (HCO_3^-) as well as carbonate (CO_3^{2-}) ions, in turn leading to declining pH values. During MARE II, initial pH values decreased during the first period of the experiment, reaching low values. This was probably due to the CO_2 excretion caused by the respiration of the cultured fish. To compensate this process, the system was supplemented with calcium oxide (CaO) in order to stabilize pH values. After experimental day 90, an increase of pH was recorded, probably due to enhanced denitrification (resulting in an increase of hydroxyl ions) within the sediment of the worm bioreactor. To counteract this process a water exchange of 1000 L had to be performed and additionally acid was given to the system to stabilize the pH. Acid supplementation was also considered to be necessary due to the increased concentrations of unionised ammonia influenced by temperature and pH-values. Alkalinity of the system water was low indicated by the rapid changes of pH even with low dosages of CaO or acid. Due to the regular dosage of CaO the formation of $Ca(HCO_3)_2$ was enabled and the newly formed $Ca(HCO_3)_2$ could be removed from the system by foam fractionaton. This could severely lower the alkalinity of the system because for each calcium (Ca^{2+}) ion two bicarbonate (HCO_3^-) ions will be removed. Future research should focus on the carbonate system and the alkalinity of recirculated water to secure a balanced carbonate system of the water.

The apparent worm growth during the MARE I experiment can be regarded as poor. High numbers of small worms were detected during the weighing processes. The total organic content remained constant at low levels indicating limited amounts of food for *N. diversicolor*. Consequently, we assumed that the worms had been feeding on smaller individuals to meet their metabolic requirements and simultaneously abundances constantly decreased throughout the experiment. At the end of the experiment, with increasing fish biomass and subsequently increased amounts of particulate matter settling in the worm tank, growth improved indicating an under-supply of energy and/or nutrients for the worms. During MARE II the total organic content of the sediment as well as the daily amount of food available for the worms was higher compared to the first experiment. This was caused by the increased fish biomass and the resulting increased amounts of fish feed supplied to the fish. Growth rates of the worms, calculated with the median, reached 0.058 d^{-1}. Due to the weight class distribution of the collected worms it was necessary to apply the median to calculate the specific growth rates. Growth rates obtained from literature are generally calculated applying the arithmetic mean. The growth rates, applying the median (μ_{max} = 0.058 d^{-1}), exceed the growth rates calculated with the mean (μ_{max} = 0.025 d^{-1}). Both results were comparable to results of Nielsen *et al.* (1995), Costa *et al.* (2000) and Bischoff (2003) but we assume that under the applied nutritional conditions growth rates reaching almost 0.06 d^{-1} are not realistic (see chapter 3). During the growth period of the worms the performance of the *Nereis* bioreactor as a sink for solid particles produced by the fish can be regarded as sufficient, due to improved growth performances of the worms and constant TOM contents of the sediment. However, as soon as the worm population reproduced, organic matter accumulated resulting in anoxic conditions which were in turn endangering the successful cultivation of the fish. Thus, the natural life cycle of an organism with essential importance for the system stability may lead to the breakdown of the system. A possible alternative in order to avoid the problems observed during MARE II, concerning the integration of *N. diversicolor*, may be the integration of several detritivorous tanks. These tanks should be equipped with *N. diversicolor* at different life cycle stages and their controlled connection to the main system can ensure the utilisation of the excreted particulate matter.

The macroalgae filter during MARE I was sufficient to ensure low concentrations of nitrite and ammonia in the recirculation water. However, growth performance of the macroalgae was inadequate and may be explained by the following reasons: the low concentrations of ammonia often only met the minimum physiological requirements of the algae but probably were too low to allow significant growth. Baird and Middleton (2004) constituted that an autotroph's ability to grow will depend on its requirement of light to fix carbon (C) and nitrogen (N), as determined by the C/N-ratio, and the rate at which it receives light and N from the environment. The artificial illumination applied during the experiments resulted in lower nutrient uptake rates compared to natural light. Like the majority of higher plants, *S. chordalis* is known to switch to nitrate uptake, when ammonia is not available as nitrogen source for its metabolism. Concurrent uptake of ammonia and nitrate has also been reported for macroalgae (Ahn *et al.* 1998). In the recirculating system, ammonia was always present at low levels caused by fish metabolism. Measurements clearly indicated that *Solieria* did not switch to nitrate uptake. Development of epiphytes was observed as a consequence of the poor growth and insufficient stocking densities of macroalgae at the beginning of the MARE I experiment. A smaller unit for the cultivation of macroalgae might have yielded better results with respect to these conditions. Another explanation for the insufficient performance of the macroalgae unit could be limitation by micronutrients, although it is assumed that sufficient amounts of micronutrients are provided to the system by fish feed (Metaxa *et al.* 2006). Nevertheless, it might be possible that the application of ozone had a negative effect on some essential ions (e.g. oxidation of Fe^{3+} to Fe^{2+}). Thus, some essential ions may not have been available for the macroalgae. An accumulation of trace elements such as copper (Cu), zinc (Zn), chromium (Cr) or cadmium (Cd) could be another explanation for the reduced macroalgal growth. Unfortunately, heavy metal concentrations within the MARE-system were not recorded, but according to Metaxa *et al.* (2006) concentrations of chromium, manganese, cobalt, nickel, copper, arsenic and thallium were significantly higher in the muscle of fish from a recirculating aquaculture system (RAS) compared to a flow through system (FTS). We assumed that increased concentrations of heavy metals in fish from a RAS indicate that higher concentrations of these heavy metals are also available for macroalgae within a RAS. Lee and Wang (2001) demonstrated that an increase in ambient nitrate concentration resulted in a significant increase in Cd accumulation and higher Zn uptake rates. Cr accumulation increased significantly with increasing phosphate concentrations. Nitrate and phosphate concentrations observed during MARE I had been in the range mentioned by Lee and Wang (2001). Thus heavy metals may have been accumulated within the macroalgae. Andrade et al. (2004) concluded that *Enteromorpha flexuosa* could not avoid penetration of Cu into the cytoplasm and its toxic effects at concentrations above 250 µg Cu L^{-1}.

The microalgae used as nutrient removal step in MARE II exhibited a completely different behaviour compared to the macroalgae used in MARE I concerning the nitrogen uptake. Although ammonia is known to be the preferred form of nitrogen and readily taken up by phytoplankton (Collos and Slawyk 1981; Levasseur et al. 1993), measurements showed that the cultivated *Nannochloropsis* sp. in MARE II mainly took up nitrate due to the lack of sufficient ammonia concentrations in the inflow water. The low flow rates through the photobioreactor system may have been the reason for an insufficient ammonia supply of *Nannochloropsis* sp. Flow rates were limited by the maximum specific growth rate and consequently growth rates did not exceed 0.025 h^{-1}. Thus the maximum dilution rate in order to avoid biomass washout had to be 0.025 h^{-1} and therefore the total maximum water flow through the entire photobioreactor system was recorded at 108 L d^{-1}, which is unfortunately a negligible volume regarding the total nutrient budget of the MARE-system. Hence, the presented dimensions of the photobioreactor system cannot fulfil the requirements of a biofilter removing the complete ammonia and nitrite concentrations from the recirculation water.

The conceptional design of the photobioreactor system appears to be adequate with respect to water pre-treatment and harvesting process. Cultivated microalgae could be used as valuable feed (e.g. rich in polyunsaturated fatty acids) for feeding organisms such as *Brachionus* sp., copepods, bivalves, *N. diversicolor* or fish larvae (Støttrup and McEvoy 2002). The fatty acids arachidonic acid (AA) and eicosapentaenoic acid (EPA), detected during the analyses of the cultured microalgae, represent two polyunsaturated fatty acids which are essential for larviculture e.g. to ensure successful survival, growth and metamorphosis of the larvae. Live feeds normally used for larval culture lack n-3 HUFAs (Highly unsaturated fatty acids) such as EPA and have to be enriched. The major source, so far for n – 3 HUFA supplementations has been fish oils whose content can vary substantially (Sargent et al. 1999). Another source for supplementation could be cultured microalgae such as *Nannochloropsis* sp.

However, possibilities for optimization can be seen: the results of both experiments indicate that the simultaneous integration of microalgae and macroalgae production may be possible, because the different algae do not compete for the nitrogen source due to the macroalgae´s preference for ammonia and the nitrate uptake of the microalgae.

The MARE experiments showed substantial insights into the microbial nitrogen cycle within the system: it seems obvious that one of the major microbial processes within the system was the denitrification process, mainly located in the sediment of the detritivorous reactor.

Tab. 8: Modelled nutrient budget obtained from the MARE-Model and the MARIS-Model (according to Wecker 2006). Positive values indicate nutrient uptake, negative values indicate nutrient release.

	MARE Nitrogen [%]	MARIS Nitrogen [%]	MARE Phosphate [%]	MARIS Phosphate [%]
S. aurata	30	26	43	37
N. diversicolor	-1	1	-1	0
S. chordalis	17	45	13	36
foam fractionation	4	4	9	10
Bacteria	3	1	5	2
Denitrification	35	12		
effluent water	6		17	
Total	94	89	86	85
Input	100	100	100	100
Difference	6	11	14	15

Tab. 8 presents the results of two newly developed models. The MARE-Model (for further details see Wecker 2006) reflects the modelled nutrient budget of MARE I.

The modelling was principally based on 15 key processes, utilising or transforming the nutrients nitrogen and phosphorous, which occur in the MARE-system. The MARIS-Model (for further details see von Harlem 2006) reflects the model output concerning an optimized system design. The optimisation process of the integrated recirculating system was based on the additional revenues of the secondary production units.

The results of the MARE-Model show that during MARE I denitrification was one of the dominant processes within the nitrogen cycle of the recirculating system. It was assumed that bacterial degradation of organic matter created anaerobic conditions within the sediment of the detritivorous bioreactor. The observed nitrate concentrations (60 – 80 mg L^{-1}) and the TOM concentration in the sediments favoured denitrification. During MARE I nitrate-N concentrations decreased and low organic contents within the sediment were recorded, indicating a balance between nitrification and denitrification. During MARE II total organic matter increased due to increased settling rates of solids from the fish and suboxic to anoxic conditions (favouring denitrification) within the sediment developed, indicated by dark-coloured sediment, absence of worms and increased occurrence of gas bubbles. Increasing P/N-ratios generally are considered to be an indication for anaerobic conditions (Wecker 2002; 2006), but in the presented system, this increase is probably caused by the subsequent addition of phosphate via feed. Evidence for an enhanced denitrification activity was obtained by i) a rapid decrease of nitrate and ii) a subsequent increase of pH as a consequence of the denitrification process (Rheinheimer et al. 1988). Additionally, bacterial mats identified as Beggiatoa sp. were observed. Beggiatoa sp. is known as a sulphur-oxidising bacterium (Madigan et al. 2001). The observed black colour of the sediment indicates the presence of H_2S, explaining the presence of Beggiatoa sp. Furthermore, the ability for denitrification is also described for Beggiatoa sp. on freshwater sediments (Sweerts et al. 1990) and indications were found for denitrification ability of this species in marine habitats as well (McHatton et al. 1996). Another nitrate-consuming process, the assimilatory nitrate reduction, is of possible relevance within the integrated system. Many bacterial species are able to reduce nitrate in order to obtain ammonia for biomass synthesis (Madigan et al. 2001). Assimilatory nitrate reduction would also lead to nitrate consumption and the two processes may be considered as major reasons for the observed decreasing nitrate values. Additionally, nitrate ammonification is performed by several bacteria in order to obtain reduction equivalents for fermentation. However, this process is inhibited by elevated concentrations of ammonia and may therefore be of minor relevance for the explanation of the observed effects.

5. Conclusions

MARE is, due to its design i.e. a land based marine recirculating aquaculture system, applying nutrient recycling as water treatment and consequently reducing drastically the requirements for water replacement, an innovative concept for new marine cultivation systems which allows simultaneously water reprocessing and nutrient recycling with the aim of enhancing economical profitability and increasing system's ecological sustainability. MARE fulfilled the demands of a closed recirculation system (< 10% water exchange per day) (Losordo et al. 1999). The daily water replacement rate due to evaporation, foam fractionation and losses by maintenance or biomass determination averaged 0.8% of the total system volume for both experimental periods.

Although the first attempts of this new type of recirculation system were beyond optimal configuration, the experimental data can help to calculate the nutrient fluxes and module sizes in virtual simulations to improve the biological and economical efficiency of future commercial recirculation systems (von Harlem 2006). The experimental outcome confirmed that the accurate adaptation of all components is crucial (Losordo et al. 1999; Waller et al. 2002). Monitoring the different nitrogen-conversion processes is strongly recommended. Therefore future experiments should simultaneously obtain oxygen concentrations in different compartments of the system and in different regions of the compartments (water, different sediment layers) in order to be able to react before anoxic conditions can develop. Although nutrient retention by secondary production accounted only for one fifth of the total nutrient budget, the financial benefit of additional harvestable biomass can significantly improve the economical outcome of an aquaculture system.

Chapter 3

The detritivorous polychaete *Nereis diversicolor* (O.F. Mueller, 1776) cultured with solid waste from recirculating aquaculture systems

Bischoff A.A., Kube N., Wecker B. und Waller U.

Abstract

Nereis diversicolor proved to be an appropriate candidate for integrated aquaculture, and its culture is one promising possibility to reduce the amount of solid waste originating from fish and/or crustacean culture within recirculating systems.

Several experiments were performed in different culture systems concerning the culture of *N. diversicolor* with solid waste as exclusive food source. Questions concerning the feasibility of worm culture with solid waste, the impact of the type of sediment on the survival and growth of *N. diversicolor*, the best growth achievable under the applied conditions, the lifecycle of *N. diversicolor* and the total organic matter content of the sediment as an indicator for the consumption of solid waste by *N. diversicolor* were addressed. Therefore, experiments in batch culture and in recirculating systems of different scales (time and space) were performed.

The type and amount of food influences the ammonia concentrations and thus, the survival of *N. diversicolor*. To avoid excess mortality rates, ammonia concentrations below 8 mg L^{-1} are recommended. The type of sediment used for worm culture affects their survival and growth. Fine grained sediment represents the best choice for the culture of *N. diversicolor*. Survival rates of 90% could be achieved. Observed growth of *N. diversicolor* was quite different, ranging from negative growth rates up to 0.025 d^{-1} applying the arithmetic mean weight of the worms. Reproduction within the applied recirculating systems could be repeatedly observed and a complete lifecycle could be achieved in approx. 110 days.

An upgrade of, otherwise discharged, nutrients by the culture of *N. diversicolor* can be achieved.

1. Introduction

During the last few years finfish and shellfish aquaculture production has exceeded 40 million tonnes per year. One of the major problems of finfish and shellfish aquaculture is the effluents loaded with dissolved and particulate nutrients. The nutrient retention of fish for feed nitrogen (N) and phosphorous (P) ranges between 20 – 50% and 15 – 65%, respectively. The fractions for nitrogen and phosphorous excretion are in the range between 32 – 64% (N) and 0 – 38% (P) and for faecal loss between 9 – 32% (N) and 33 – 63% (P) (Schneider et al. 2005).

Different production systems, such as open and closed systems, are applied for aquaculture. Open aquaculture refers to production systems, which interact with the surrounding environment whereas closed aquaculture systems do not interact. Differing nutrient loads released by open or closed aquaculture causes effects on diverse scales. For open aquaculture systems, e.g. net cages, dissolved and particulated nutrients are rapidly removed by surrounding waters. Dissolved nutrients, which are fertilising the surrounding waters, are able to enhance primary production, leading to increased algal biomass which can lead to oxygen depletion by their degradation or in severe cases to outbreaks of harmful or fishkilling algal blooms (Kim 1997; Rabalais 2002; Islam et al. 2004). Organic enrichment of sediments below aquaculture sites, caused by sedimentation of particulate nutrients, leads to environmental problems ranging from benthic community shifts to azoic conditions (Pearson and Rosenberg 1978).

In closed recirculating aquaculture systems dissolved and particulate nutrients may lead to similar problems. Increases of inorganic dissolved nutrients such as ammonia (NH_3) or nitrite (NO_2^-) may lead to behavioural changes of cultured species or with further increasing concentrations to death of the animals (Tilak et al. 2002; Fuller et al. 2003; Lemarie et al. 2004; Weirich and Riche 2006). Suspended solids in the water of closed recirculating systems may cause Bacterial Gill Disease (BGD) by cultured fish (Bullock et al. 1994). Suspended solids can act as microhabitats and inhabit bacteria on the surfaces and thereby distribute bacteria within the entire system. This could cause higher oxygen consumption by bacteria within the water and consequently decreases the amount of available oxygen for fish (Curds 1982; Heissenberger et al. 1996; Acinas et al. 1999; Kube and Rosenthal 2006). The increase of suspended solids and as a result of the organic load within the water will inhibit the nitrification process (Bovendeur 1989).

Due to environmental problems and problems for the cultured species caused by nutrient loaded waters, more attention was given to the sustainable development of aquaculture during the last few years. The fact that cultured species use only about 30% of the supplied nitrogen, discharging the rest into the environment and polluting thereby both freshwater and marine habitats, is still quite common. Predictions concerning the world fish supply for the next decades forecasts an increase in aquaculture production from 42.3 million tonnes in 2003 (aquatic plants excluded) up to 59.7 million tonnes for 2010 and 209.5 million tonnes for 2050 (Wijkstrom 2003). These forecasts are based on the assumption that capture fisheries will be stagnant or even declining but fishmeal supply will be constant over time. The first assumption seems to be reasonable but for the second one there are legitimate doubts (Hardy and Tacon 2002). Besides aquaculture production also terrestrial animal production depends on fishmeal as a cheap and nutrient rich ingredient for formulated animal feeds. Therefore, it can be assumed that fishmeal supply will change in the future. First indications for this expected scenario already occurred, noticeable at the price development of fishmeal during the last few years. At the end of 2005 prices per tonne fishmeal were around US$ 550 and have risen to above US$ 1000 in May 2006, reaching maximum values of more than US$ 1300, with prices to be expected to stabilize at levels of US$ 1050 per tonne (FAO 2006b).

One possible solution for the listed problems above might be the concept of integrated aquaculture as it was described by Neori *et al.* (2000), Chopin *et al.* (2001) or Schneider *et al.* (2005). The concept combines the culture of fish or crustaceans as the primary organism of an aquaculture production with further organisms belonging to lower trophic levels and using the waste products of the primary organisms as resource for their own production. Primary and secondary production refer to the production realized by the use of commercial diets (primary production) or the additional production (secondary production) of organisms lower in the food chain, which uses the remaining nutrients in the waste excreted during the fish and/or crustacean production. Combinations of primary producers (algae), filter feeders (molluscs), herbivorous (sea urchins) and detritivorous organisms (worms) with fish are described in more detail by Schneider and co-authors (2005). The authors state the gain of efficiency by the combination of these organisms. Von Halem (2006) was the first study, which determined the required space for optimum secondary production of an integrated recirculating system in a modelling approach. Results indicate that an additional economical benefit of 10% is achievable for an aquaculture production site (100 tonnes fish per annum) by culturing and selling detritivorous worms. Nevertheless, little information is available about the culture of detritivorous invertebrates as consumer of particulated wastes from aquaculture. Batista et al. (2003) reported about trials with Nereis diversicolor to use faeces of the carpet shell clam Ruditapes decussatus (L., 1758) as food source. Besides that little information was published on the use of detritivorous organisms in integrated aquaculture systems. According to different authors about 10 to 15% of the given food is lost as faecal material during commercial aquaculture production (Schneider et al. 2005). This particulate waste represents a valuable source of e.g. carbon, nitrogen and phosphorous, which can be used by different organisms.

The common ragworm N. diversicolor gained during the last few years importance from ecological as well as aquacultural side. From the ecological point of view questions concerning the use of N. diversicolor as indicator for cadmium, copper, zinc and other pollutants in its benthic environment were of interest (Elder et al. 1979; Berthet et al. 2003; Frangipane et al. 2005). Dhainaut and co-authors (1989) reported about an antibacterial protein in the celomic fluid of the worms. Hansen and Kristensen (1997) as well as Heilskov and Holmer (2001) focussed on the worms influence on benthic metabolism and organic matter mineralization in the vicinity of net cages. In addition, polychaetes represent high quality live feeds for finfish and crustacean aquaculture (Luis and Passos 1995; Costa 1999) which stimulate gonad maturation and influence the fry quality.

The use of solid waste by N. diversicolor as a food source seems obvious and was tested by Batista et al. (2003) for faeces from the carpet shell clam R. decussatus. They suggested that faeces could act as a substrate for bacteria and protozoa which could be ingested by N. diversicolor. This statement is supported by Montagna (1984) who showed that about 3 % h^{-1} of the present bacteria community within the sediment was removed. This removal process was mainly dominated by polychaetes. Furthermore, Lucas and Bertru (1997) demonstrated that N. diversicolor is able to degrade bacteria in its digestive tract and to obtain therefore a nutrient gain.

During feasibility studies, evaluating the ability of N. diversicolor to grow on a diet of uneaten fish feed and fish faces (Bischoff 2003), elevated mortality rates were observed. Mortality up to 37% for fed worms was recorded. These high mortality rates were in contrast to results about the culture of N. diversicolor in different sediments and differing diets (Costa et al. 2000). Obtained specific growth rates during the study of Bischoff (2003) were lower compared to results of Nielsen et al. (1995), Riisgard et al. (1996) and Costa et al. (2000). Organic enrichment and anaerobic conditions are developing in the close vicinity of aquaculture production sites (Heilskov and Homer 2001). Saiz-Salinas and Frances-Zubillaga (1997) observed that with improving oxygen conditions organic enriched sediments can be re-populated by juvenile N. diversicolor, showing increased growth even without the addition of exogenous food. The worms used the nutritional quality of the applied sediments, which were assumed to be the organic matter of the sediment.

The taxonomic classification in addition to a short biological and ecological description combined by Bischoff (2003) is displayed in Tab. 1.

The focus of this research was to investigate the biology and ecology of N. diversicolor in the context of aquacultural purposes. The potential of particulated waste originating from recirculating aquaculture systems used as a food source by N. diversicolor was examined. Therefore, the biological requirements of the worms such as nutritional demands, resistance to elevated nutrient concentrations, as well as the impacts of sediment conditions were altered under cultural conditions in so called bioreactors. As indicators for the performance of the bioreactors the survival and growth of the worms as well as the expected utilisation of applied solid waste by the worms were recorded. As a proxy for the utilisation of the solid waste, the organic content of the sediments was recorded.

Tab. 1: Taxonomic classification, biological and ecological description of *N. diversicolor* combined by Bischoff (2003) according to Hartmann-Schröder (1996) and Scaps (2002).

Phylum:	Annelida
Class:	Polychaeta
Order:	Phyllodocida
Family:	Nereididae
Genus:	Nereis
Synonyms	*Hediste diversicolor*, common rag worm, red rag worm
Max. size	Up to 20 cm, separated in max. 120 segments
Environment	Infaunal species; inhabits tubes on sandy muds, gravels, clay and even turf
Climate	Able to tolerate great changes in temperature, ranging from +4 to +25 °C
Distribution	Shallow marine and brackish waters on the coasts of the North temperate Atlantic Ocean
Biology and Ecology	*Nereis diversicolor* is euryoecious and euryhaline, which means it is able to adapt to a wide variety of environmental conditions, including drastic hypoxic and saline conditions. The common rag worm is able to adapt its feeding behaviour to the environmental conditions, switching between omnivorous, planktivorous, bacteriovorous, herbivorous and carnivorous. Reproduction is monotelic that means it reproduces only once in it's live, reproduction is always followed by death. During maturation, individuals change from a slight yellow/reddish-brown colour to bright green (male) and dark green (female). Individuals inhibited to reproduce elongate their somatic growth for a certain period.

The following specific questions concerning *N. diversicolor* were addressed:

- Is it possible to culture *N. diversicolor* on an exclusive diet of solid waste from fish culture?
- Which impact has the type of sediment on the survival of *N. diversicolor*?
- What influence has the type of sediment on the growth of *N. diversicolor*?
- What is the best growth achievable under the applied conditions?
- Are the applied conditions adequate to complete a lifecycle of *N. diversicolor*?
- Is the total organic matter content of the sediment a reliable indicator for the consumption of solid waste by *N. diversicolor*?

2. Material and Methods

2.1 Nereis diversicolor

Worms used for the experiments and as potential brood stock in the laboratory were collected at different dates at the Atlantic coast of France, distributed by TopsyBaits (www.topsybaits.com) and delivered via The Netherlands to Kiel, Germany. After the arrival at the Leibniz-Institute of Marine Sciences in Kiel the worms were acclimatised in their transport bags for several hours and transferred afterwards into holding tanks which were part of a recirculating system (salinity approx. 30 psu, temperature 12 – 14 °C).

All holding tanks were filled with a five to ten centimetre deep sand layer four weeks prior to the stocking with worms. The sediment functioned as substrate for the worms. Each individual tank was equipped with an air supply to secure sufficient oxygen saturation in the water. Water was recirculated at approx. one tank volume per hour. Water treatment was done by an aerobic nitrifying filter transforming excreted ammonia into nitrate.

During the 48 hours following the stocking of the worms, all dead individuals (caused by transportation and handling) were removed from the tanks to keep water quality as good as possible.

2.2 Experimental set up

To answer the questions concerning the consumption of solids by N. diversicolor as well as the survival and growth of the worms, different set ups were used to investigate these topics. Tab. 2 compares the different experimental systems in respect to numbers of treatments, replicates as well as water and sediment volumes.

Tab. 2: Comparison of the number of treatments and replicates as well as water and sediment volumes applied during the different experiments with N. diversicolor.

Culture system	Number of treatments	Number of replicates	Total number of tanks	Water volume $[L\ tank^{-1}]$	Sediment volume $[m^3\ tank^{-1}]$
Batch	4	3	12	6	0.004
Recirculating system					
Small scale (preference)	6	4	24	2.5	0.001
Small scale (dissolved)	4	3	12	6	0.004
Medium scale	2	3	6	210	0.050
Multitrophic integrated (MARE)	1	1*	1	1350	0.250

*Two experiments were performed for MARE, therefore one was assessed as replicate.

2.2.1 Survival

2.2.1.1 Is it possible to culture *N. diversicolor* on an exclusive diet of solid waste from fish culture?

Experiments investigating simultaneously the survival and growth of the worms were carried out. During the experiments, dissolved nutrients such as Total Ammonia Nitrogen (TAN), nitrite-N (NO_2-N), nitrate-N (NO_3-N) and orthophosphate-P (PO_4^{3-}-P) were analysed on a daily basis. Dissolved inorganic nutrients were analysed using photometric test kits (Hach Lange, Düsseldorf, Germany) or continuous-flow wet-chemical analysis (Bran+Luebbe, Norderstedt, Germany). For both procedures analytical methods for dissolved inorganic nutrients were identical. Total Ammonia Nitrogen (TAN) was analysed using Nesslers reagent, the indophenol blue reaction and the salicylate method. Nitrite concentrations were determined using the diazotation method. Nitrate concentrations were analysed using the cadmium or hydrazine reduction method and orthophosphate was measured using the amino acid method (for more details see Grasshoff *et al.* 1999). Abiotic parameters, such as water temperature, salinity, oxygen saturation and pH were also recorded on a daily basis. Simultaneously, the observed mortality was recorded daily, by counting and collecting dead individuals from the sediment surface of the tanks. Total mortality was calculated as the difference of stocked and recaptured individuals. In cases were all individuals were used for biomass determination, the total mortality was employed. In the case of the medium scale recirculating system total mortality was not applicable and therefore, observed mortality was employed.

To investigate the influence of supplied food on the survival of *N. diversicolor* three different set ups were used:

2.2.1.1.1 Batch culture

For the batch culture experiment a total of twelve individual tanks (each 36.0 x 19.3 x 20.8 cm; total tank volume 12 L), which had no connection to each other, were stocked with 35 worms per tank. Each tank was applied with an individual air supply. Applied sediment was fine sand (grain size ≤ 2 mm) and food source was commercial fish feed (Biomar *Aqualife 17;* for biochemical composition see Tab. 6). Treatments 1 to 3 were fed with the commercial fish feed Biomar *Aqualife* 17 (for feeding rates see

Tab. 4). Treatment 4 did not receive any food at all. During the experimental period each week a partial water exchange was performed (exchanged water volume per week: 6 L tank^{-1}).

2.2.1.1.2 Small scale recirculating system

A total number of twelve individual tanks (each 36.0 x 19.3 x 20.8 cm; total tank volume 12 L) were used to build four small scale recirculating systems with three tanks each. These systems will be referred to as small scale recirculating system (dissolved). Each individual system was equipped with a submersed biofiltration unit and a pump. Applied sediment was fine sand (grain size ≤ 2 mm). Each individual tank was stocked with 35 worms. Treatments 1 to 3 were fed with solid waste collected from a recirculating system for the cultivation of sea bass (*Dicentrarchus labrax*) at different feeding rates (for feeding rates see Tab. 4; for biochemical composition of solid waste see Tab. 6) and treatment 4 was fed with commercial fish feed (Biomar *Aqualife 17*; see Tabs. 4 and 6).

2.2.1.1.3 Medium scale recirculating system

Six individual tanks (0.98 x 0.48 x 0.44 m; total tank volume 210 L) were connected to one recirculating system which was also equipped with a trickling biofilter, a foam fractionator and a pump. Three tanks were stocked with approx. 600 worms each and the other three tanks were used as a control without worms. Applied sediment for all tanks was fine sand (grain size ≤ 2 mm). All six tanks were fed the same amount of solid waste, which was collected from a recirculating system for the cultivation of sea bass (*D. labrax*). The amount of solid waste was calculated to supply each worm with a maximum of 1 kJ d^{-1} (see also Tab. 4).

2.2.1.2 Impacts of sediment on survival

To investigate the influence of the sediment on the survival of the worms a total of 24 small tanks (each 19.5 x 13.2 x 12.0 cm; total tank volume 2.5 L) were combined with a trickling biofilter and a pump to form a recirculating system, which will be referred to as small scale recirculating system (preference). Four tanks were used for each of the six treatments (Sed. 1 – Sed. 6). The tanks of the different treatments were equipped with diverse substrates including fine and coarse sand (Sed. 1: grain size ≤ 2 mm; Sed. 2: grain size 2 – 4 mm), ceramic filter media for aerobic nitrifying filters (Sed. 3), one set of tanks without any substrate at all (Sed. 4), PVC mats (Sed. 5) and Aquaclay® (Sed. 6: grain size 3 – 5 mm). All supplied substrates had a depth of 3 to 5 cm, except Sed. 4 (no substrate at all).

Each tank was stocked with ten worms. Applied food for all tanks was solid waste collected from a recirculating system for the cultivation of sea bass (*D. labrax*). Each tank was supplied with about 3% wet weight of the stocked worm biomass (for feeding rates see Tab. 4). Simultaneously, observed dead individuals at the sediment surface of the tanks were collected each day.

2.2.2 Growth

To investigate the growth of *N. diversicolor*, results obtained from different culture systems were used. For systems description of the small scale and medium scale recirculating systems see sections 2.2.1.1.2 and 2.2.1.1.3. For systems description used for the investigations concerning growth of the worms in an integrated multitrophic recirculating system see Chapter 2.

Biomass determination was done according to Bischoff (2003). Worms were dried on paper tissue prior to weighing. Average individual bodyweight was calculated by using only complete individuals. The specific growth rate μ of *N. diversicolor* for the different experiments was calculated according to Jørgensen (1990) using the equation:

$$\mu = ln(W_t / W_o) * t^{-1} \qquad (1)$$

where W_0 and W_t are the average body mass (wet weight) of the polychaetes on Day 0 and Day t respectively.

2.2.3 Consumption of solid waste by *N. diversicolor*

Total organic matter of the sediment was used as indicator for the consumption of supplied food by the worms. Each tank from the small scale recirculating system (dissolved) (section 2.2.1.1.2), the medium scale recirculating system (section 2.2.1.1.3) and the integrated multitrophic recirculating system (Chapter 2) was sampled at least on a weekly basis to determine the total organic matter content of the sediment. For the location of the different sampling points for the multitrophic integrated recirculating system (MARE I and II) see Chapter 2, section 2.2.2 Solid components. Analyses were performed according to Winberg and Duncan (1971). Each sample was dehydrated in a drying furnace at 60 ± 5 °C for 24 hours. Samples were weighed prior to incineration in a muffle furnace at 450 ± 50 °C for 24 hours. After cooling down samples were weighed again and the difference was used to calculate the TOM content of the sediment.

2.3 General experimental considerations

Tab. 2 summarizes and compares the different culture systems emphasizing on the number of treatments and replicates, water volume of each individual tank and the sediment volume applied during the different experiments.

Tab. 3 expresses the number of individuals per tank, the sediment surface area per tank in m^2 and the calculated stocking density per m^2 for the different applied culture systems used during the experiments with *N. diversicolor*.

Tab. 4 combines all food sources used during the experiments. The different feeding groups applied during the experiments as well as the feeding regimes are displayed.

Tab. 5 presents the duration of all accomplished experiments. Duration of the different experiments varied from short term (ten days) to long term (five months).

Tab. 3: Number of individuals per tank and calculated stocking density per m^2 for the different culture systems. Stocking density was calculated using the sediment surface area and number of individuals of the according tanks.

Culture system	Number of individuals [$tank^{-1}$]	Sediment surface area [$m^2\ tank^{-1}$]	Stocking density [m^{-2}]
Batch	35	0.07	≈ 950
Recirculating system			
Small scale (preference)	10	0.02	≈ 390
Small scale (dissolved)	35	0.07	≈ 950
Medium scale	500 - 600	0.47	≈ 1000
Multitrophic integrated (MARE)	1900 - 2400	2.08	≈ 950

Tab. 4: Overview of all food sources applied during the experiments: Fish feed[1] was extruded fish feed (*Biomar Aqualife 17*); Solid waste[2] was a combination of uneaten fish food and fish faeces collected from a recirculating system for European sea bass (*D. labrax*); solid waste[3] was also a combination of uneaten fish food and fish faeces of Gilthead sea bream (*S. aurata*) from the integrated multitrophic recirculating system MARE.

Culture system	Food source	Feeding groups	Feeding regime
Batch	Fish feed[1]	0, 2, 4 and 6 [% BW d^{-1}]	once day^{-1}
Recirculating system			
Small scale (preference)	Solid waste[2]	approx. 3 [% BW d^{-1}]	once day^{-1}
Small scale (dissolved)	Fish feed[1]	1 [% BW d^{-1}]	once day^{-1}
	Solid waste[2]	0.5, 1.0 and 1.5 [% BW d^{-1}]	once day^{-1}
Medium scale	Solid waste[2]	≈1 [kJ worm^{-1} d^{-1}]	once day^{-1}
Multitrophic integrated (MARE)	Solid waste[3]	≈15% [DW] of daily supplied food	continuous

Tab. 5: Duration in [d] for each experiment within the different culture systems

Culture system	Experimental duration in days
Batch	30
Recirculating system	
Small scale (preference)	10
Small scale (dissolved)	30
Medium scale	40
Multitrophic integrated (MARE)	approx. 200

Abiotic water parameters such as water temperature, pH, oxygen saturation and salinity were measured on a daily basis using handheld devices (WTW pH meter pH340, conductivity meter LF 330, Multi Parameter Instruments Multi 340i and 350i; WTW, Weilheim, Germany). Each probe was calibrated prior to measurements.

3. Results

3.1 Abiotic water parameters

Tab. 7 expresses the abiotic water parameters monitored during all experiments. Temperature was similar during all experiments ranging from 18 to 20 °C. Salinity showed more variation between different experiments but for each experiment worms were allowed to acclimatise to applied conditions for several days before the start of the experiments. pH showed minor variations and was most of the time in the range of natural seawater. Oxygen saturation of the water column above the sediment showed for all recirculating systems oxic conditions but for the batch culture experiment oxygen deficient areas were detected.

3.2 Is it possible to culture *N. diversicolor* on an exclusive diet of solid waste from fish culture?

3.2.1 Dissolved inorganic nutrient concentrations

During the batch culture experiment, investigating the amount of food necessary to achieve high survival rates of the worms, the concentrations of dissolved inorganic nutrients and in particular TAN (Total Ammonia Nitrogen) were elevated.
Fig. 1 displays the TAN concentrations obtained during the batch culture experiment. Very high TAN concentrations were recorded, depending on the applied feeding rate, exceeding values of 500 mg L^{-1} for the highest feeding rate. Figs. 2 and 3 present results from the small scale and medium scale recirculating system, respectively. Fig. 2 presents TAN concentrations for the experiment applying solid waste for the first time in a recirculating system as food source with different feeding rates.

Tab. 6: Biochemical composition (means ± SD) of fish feed and solid waste used during the experiments with *N. diversicolor*. (n ≥ 3)

Content	Fish feed (Biomar Aqualife 17)	Solids from European Seabass	Solids from Gilthead Sea Bream	[unit]
Water	~ 6.4	94.63 ± 1.31	93.15 ± 0.20	%
Organic	78.10 ± 2.93	64.57 ± 3.99	49.42 ± 0.58	%
Energy	23.36 ± 0.08	14.64 ± 0.50	9.99 ± 0.24	[kJ g_{DW}^{-1}]
Carbon	50.20 ± 1.03	36.72 ± 1.53	26.24 ± 0.33	%
Nitrogen	9.38 ± 0.72	3.45 ± 0.36	1.95 ± 0.09	%

Tab. 7: Average abiotic water parameters (means ± SD) monitored during all experiments monitored on a daily basis. (n ≥ 3)

Culture system	Temperature $_{Water}$ [°C]	pH	O_2 saturation [%]	Salinity [psu]
Batch	18.0 ± 0.5	7.91 ± 0.21	75.0 ± 32.8	30.9 ± 2.9
Recirculating system				
Small scale (preference)	19.4 ± 0.3	8.19 ± 0.07	> 70	29.5 ± 0.5
Small scale (dissolved)	19.8 ± 0.6	7.99 ± 0.20	80.6 ± 9.6	31.4 ± 1.8
Medium scale	19.4 ± 0.4	8.06 ± 0.17	71.7 ± 15.9	24.9 ± 0.2
Multitrophic integrated				
MARE I	19.9 ± 0.5	7.96 ± 0.26	74.0 ± 15.5	26.3 ± 1.4
MARE II	19.3 ± 0.5	7.29 ± 0.55	72.7 ± 8.6	24.8 ± 1.0

TAN concentrations reached high values but did not exceed 60 mg L^{-1} for the treatment feeding fish feed to the worms. The treatments which were fed solid waste showed lower concentrations compared to the fish feed treatment. Concentrations were almost half compared to the fish feed group.

Fig. 3 displays TAN concentrations of the medium scale recirculating system. TAN concentrations did not reach values as high as presented for the small scale system (dissolved). Values below 4 mg L^{-1} were recorded.

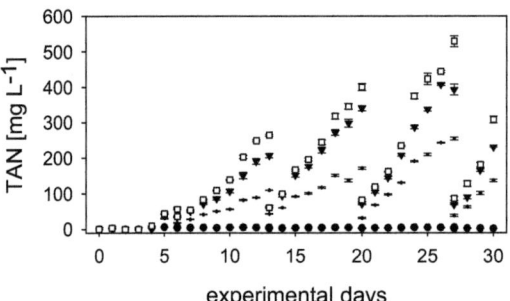

Fig. 1: TAN concentrations (means ± SE) during batch culture. Treatments were different feeding rates and are presented by different symbols: ● = 0%, + = 2%, ▼ = 4%, □ = 6% fish feed. (n = 3)

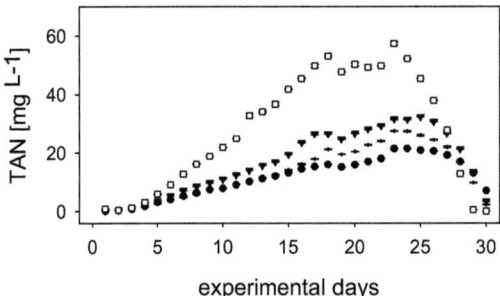

Fig. 2: TAN concentrations (means ± SE) during the culture of *N. diversicolor* in a small scale recirculating system. Treatments were different feeding rates and are presented by different symbols: ● = 0.5%, + = 1.0%, ▼ = 1.5% solid waste; □ = 1.0% fish feed. (n = 3)

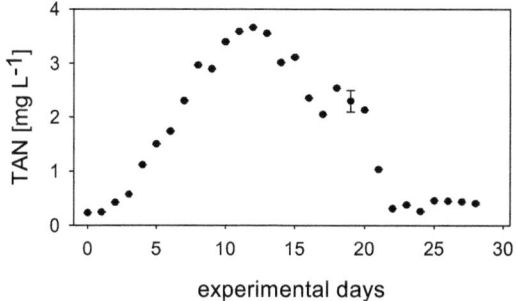

Fig. 3: TAN concentrations (means ± SE) during the culture of *N. diversicolor* in a medium scale recirculating system. (n = 3)

3.2.2 Survival of *N. diversicolor*

Simultaneously to the elevated TAN concentrations, high mortality rates of the worms were recorded during the batch culture experiment. (Fig. 4a – d). Increasing TAN concentrations in the feeding groups during batch culture deteriorated drastically the conditions necessary for the cultivation of *N. diversicolor*. Average daily mortality of the experimental groups is presented as observed mortality i.e. counting and collecting dead individuals from the sediment surface of each tank.

Mortality for all feeding groups were different compared to the starvation group, showing high mortality during the whole experimental period. Mortality ranged between 0 and 2 dead individuals per day until experimental day 10. After day 10, the mortality increased rapidly for treatment 3 (Fig. 4c), followed by treatment 2 (Fig. 4b) and treatment 1 (Fig. 4a). Cumulative mortality reached 100% for all feeding groups, except for tank 8 (treatment 1, replicate 2) where one individual survived the experiment. Mortality of the starvation group was low, showing most occurring incidents at the beginning and the end of the experiment (Fig. 4d). Survival rates for the starvation group ranged from 80% in tanks 10 and 12 up to 91% in tank 5. A correlation between mortality and TAN concentrations was assumed after the batch culture experiment (Figs. 1 and 4).

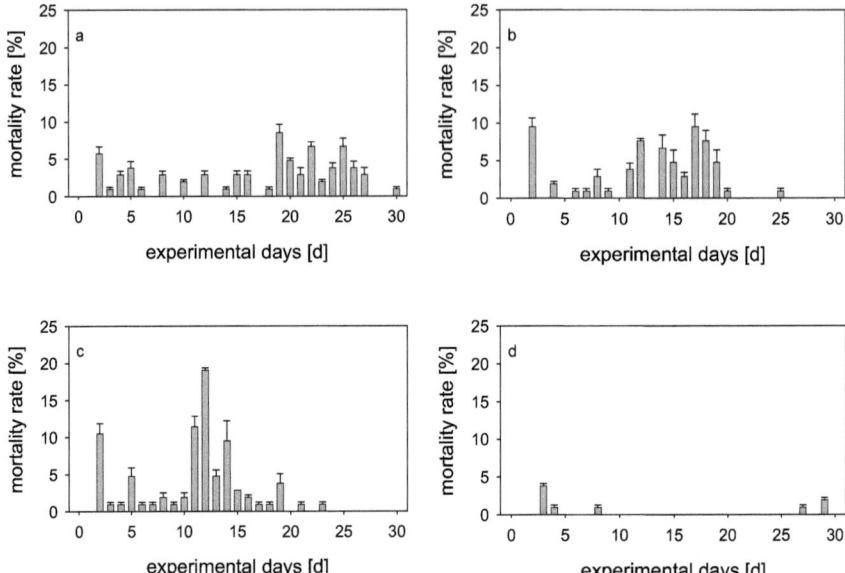

Fig. 4: Average mortality rates (means + SE) during the batch culture experiment: a) treatment 1 (2% fish feed), b) treatment 2 (4% fish feed), c) treatment 3 (6% fish feed) and d) treatment 4 (0% fish feed). (n = 3)

This assumption was further investigated during the next experiment within the small scale recirculating system (dissolved). Figs. 2 and 5 represent the recorded TAN concentrations and the mortalities obtained for the different treatments during this experiment. Absolute mortality increased with increasing TAN concentrations. Due to high TAN concentrations for the fish feed group (see Fig. 2) the survival rate for that group was zero. For the other groups survival rates ranged between 8% (faeces III, feeding rate 1.5%) and 30% (faeces I, feeding rate 0.5%).

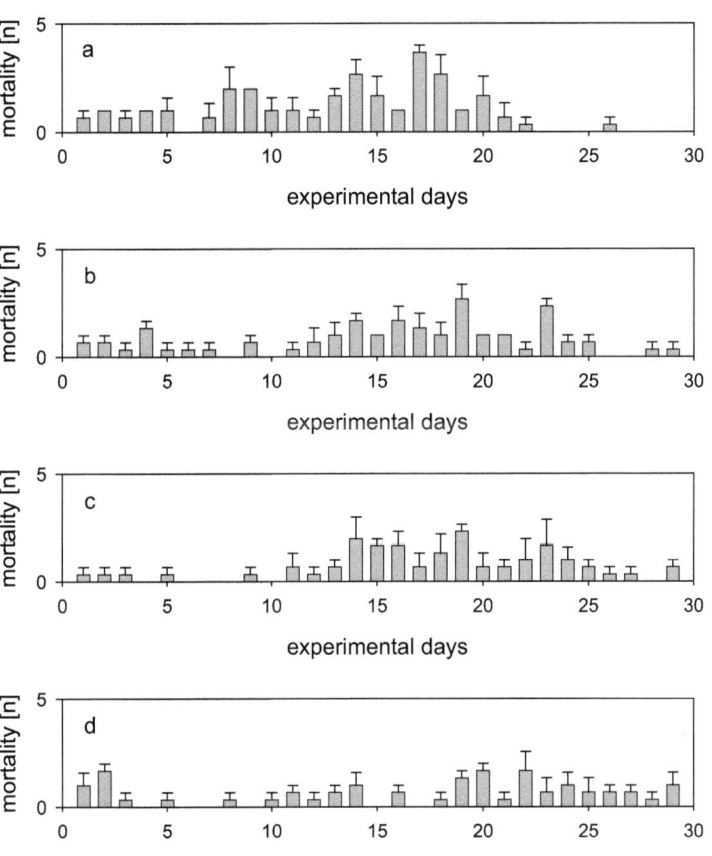

Fig. 5: Mortality obtained during the growth experiment with *N. diversicolor* within the small scale recirculating system (dissolved). Displayed values are means ± SE (n = 3). a) treatment 1.0 % fish feed; b) treatment 1.5 % faces; c) treatment 1.0 % faeces and d) treatment 0.5 % faeces.

Fig. 6 displays the observed mortality recorded during the experiment in the medium scale recirculating system. Mortality increased rapidly until experimental day 3 to a peak of 39.3 ± 10.2 individuals per day and decreased afterwards. From experimental day 7 onwards, mortality was stable at around 1 to 2 dead worms per day until day 23 where a slight increase was noticeable. After a few days mortality was stable again until experimental day 30, where mortality was increasing slightly and stabilised at a level of approx. 2 to 3 dead individuals per day.

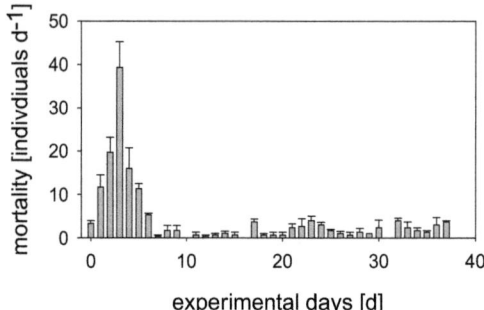

Fig. 6: Mortality rates (means + SE) of *N. diversicolor* during the experiment in the medium scale recirculating system. (n = 3)

3.3 Which impact has the type of sediment on the survival of *N. diversicolor*?

Fig. 7 gives the survival rates of the worms after the experiment for evaluating the sediment impact on *N. diversicolor*. Sediment 4 (Sed. 4) showed the lowest survival rates between 30 and 90%, with an average (± SD) of 63 ± 26%. Sediments 2 and 6 (Sed. 2 and Sed. 6) showed the highest survival rates between 80 and 100%, resulting in average survival rates (± SD) of 90 ± 8% and 90 ± 0%, respectively. No significant difference between substrates could be detected (ANOVA, p = 0.103).

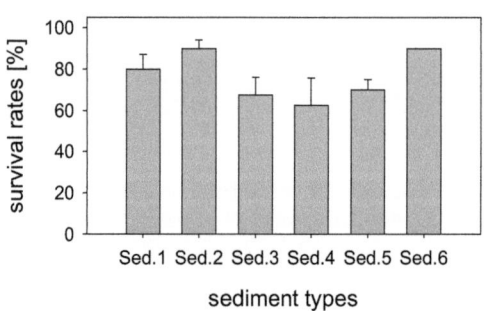

Fig. 7: Survival rates of *N. diversicolor* obtained during experiments with different substrates (means + SE) (n = 4).
The applied substrates were: Sed. 1 = fine sand; Sed. 2 = coarse sand; Sed. 3 = ceramic filter media; Sed. 4 = no substrate at all; Sed. 5 = PVC mats and Sed. 6 = Aquaclay ®.

3.4 Growth of *N. diversicolor*

In the batch culture experiment no final biomass could be determined for the feeding groups due to high mortality and therefore, no growth could be calculated. Nevertheless, survival of the starvation group was good and therefore data obtained during the experiment could be used to calculate the specific growth rates µ. Fig. 8 gives the specific growth rates µ for all three tanks. Growth rates were negative, caused by the starving of the worms. Specific growth rates μ ranged from -0.0035 d^{-1} in tank 12 to -0.0109 d^{-1} in tank 10.

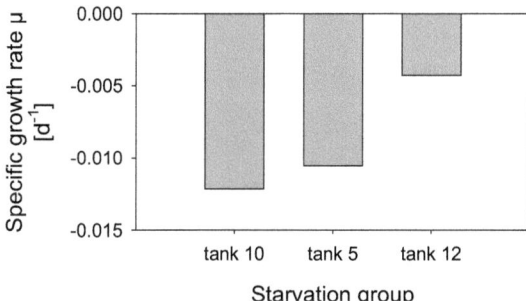

Fig. 8: Specific growth rates μ of *N. diversicolor* obtained during the batch culture experiment from the starvation group (treatment 4, 0% fish feed).

In the small scale recirculating system growth of *N. diversicolor* was used as indicator for the preferred sediment. Five different substrates plus one set of tanks without any substrate were applied. Fig. 9 presents the specific growth rates μ of the worms obtained during the sediment trials.

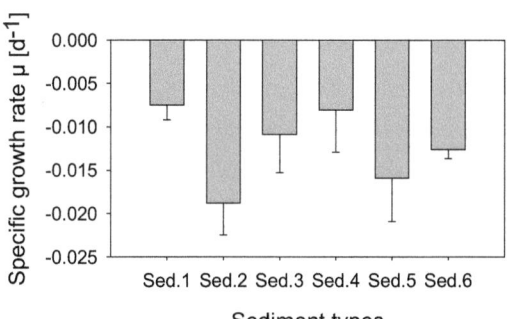

Fig. 9: Specific growth rates μ (means - SE) of *N. diversicolor* obtained during experiments with different substrates as sediments for the worms (n = 4). (For substrate declaration see Fig. 7)

For all treatments negative growth rates were recorded. Sediment 1 (Sed. 1, fine sand) presented the lowest negative specific growth rate (μ = -0.008 ± 0.002 d^{-1}), whereas, sediment 2 (Sed. 2, coarse sand) presented the highest negative specific growth rate (μ = -0.019 ± 0.004 d^{-1}). At the given variance the observed differences were statistically not significant (ANOVA, p = 0.323).

In the medium scale recirculating system the initial average bodyweight of *N. diversicolor* was calculated by collecting and weighing three sub-samples from all worms (n = 45, for each sub-sample). Average weights of the three sub-samples showed no statistical significant differences and thus, samples were pooled to calculate the average initial weight of the worms. Animals were then distributed to the tanks of either the starvation or the feeding group. After terminating the experiment, subsamples from the feeding groups and all animals from the starvation groups were collected and weighed to determine the final average bodyweight of the worms.

Fig. 10 shows the results of the average bodyweight for the three different groups (start, starvation and feeding group). At the start average weight was around 600 mg. All three tanks of the starvation group lost weight during the course of the experiment. Differences were between 34 and 183 mg and differences between the start group and the starvation groups 2 and 3 were significant (t-test, p = 0.008 and p < 0.001, respectively). No statistically significant difference between the start group and starvation tank 1 was detected (t-test, p = 0.397). All tanks of the feeding group gained weight. Differences were between 101 to 202 mg. No significant difference between the feeding groups were detected (ANOVA, p = 0.190) and therefore, the data were pooled. Difference between the start group and the feeding group was significant (t-test, p < 0.001).

Comparing the total initial (~1.05 kg) and final (~1.10 kg) biomass revealed that only a slight increase (0.05 kg) could be recorded. Taking the initial mortality into account a total biomass increase of 0.20 kg could be recorded.

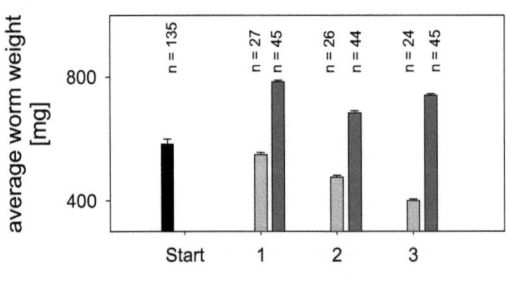

Fig. 10: Average bodyweight (means + SE) of *N. diversicolor* obtained during the experiment in the medium scale recirculating system.
— Start
— Starvation End
— Feeding End

Fig. 11 displays the specific growth rates µ calculated according to Jørgensen (1990) for all tanks used for this growth experiment. Specific growth rates µ ranged from

-0.002 to -0.010 d^{-1} for the starvation groups (Stav.1 to Stav.3) and from 0.004 to 0.008 d^{-1} calculated for the three feeding groups (Feed.1 to Feed.3).

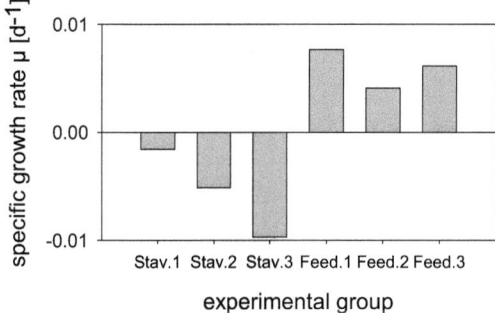

Fig. 11: Specific growth rates μ of *N. diversicolor* calculated from the average bodyweights (Fig. 10) obtained during the experiment in the medium scale recirculating system. Stav. 1 – 3 represent the starvation groups and Feed.1 – 3 represents the feeding groups.

The MARE-system was a newly designed recirculating system for evaluating the potential of integrated land based aquaculture systems for the co-culture of organisms belonging to different trophic levels. Fig. 12 presents the average bodyweight of *N. diversicolor* during the experimental stages of MARE I and MARE II (see Chapter 2). Additionally to bodyweights, the size class distributions of collected worms are presented as inlets in the graphs. During MARE I the average weight (arithmetic mean) of the worms was constant at levels close to 100 mg during most of the time of the experiment. On the last measuring day (experimental day 216) the average bodyweight increased and a shift from the smallest size class to a larger size class was noticeable.

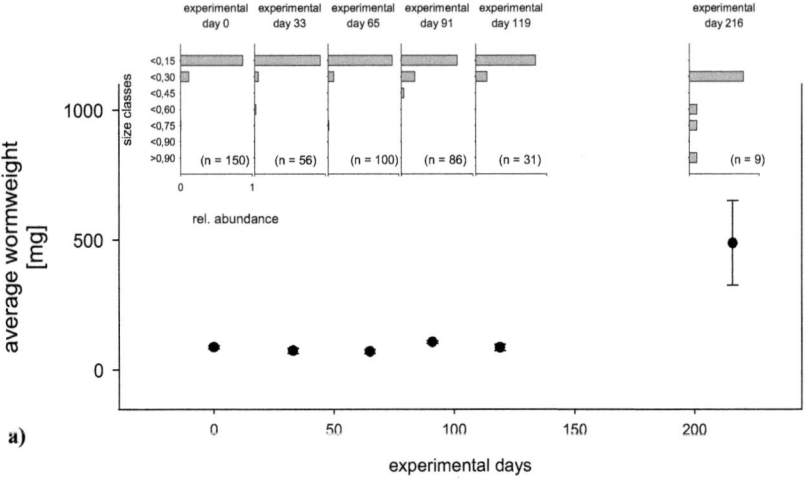

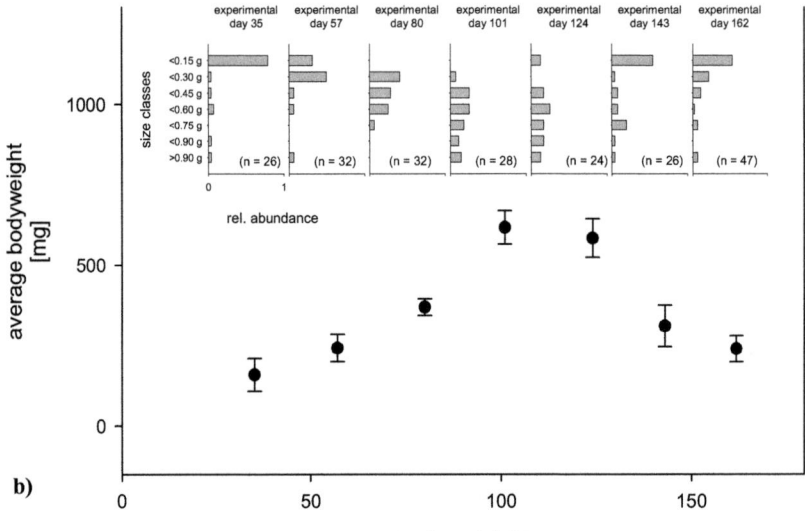

Fig. 12: Growth of *N. diversicolor* during the MARE-experiments. Main graph presents the average bodyweight ± SE; the inlet graphs represent the size class distribution of collected worms. a) results obtained during MARE I and b) results obtained during MARE II.

The growth performance of N. diversicolor during MARE II differed compared to MARE I. The average worm weight (± SE) increased constantly from an initial average value of 159.9 ± 49.0 mg at the start to 617.2 ± 51.7 mg at experimental day 101. From there on average bodyweight decreased to reach a low of 240.2 ± 40.7 mg at the end of the experiment. During the experiment it was noticeable that the size classes of N. diversicolor were moving from smaller to larger size classes until experimental day 101 and afterwards that the smallest size class was dominant again. It was visible during the ongoing experiment that after experimental day 124 the size classes were shifting again from the smallest to larger size classes.

3.5 Growth performance of N. diversicolor

Fig. 13 displays growth data of N. diversicolor obtained during the feasibility study of Bischoff (2003). These data were used for regression analysis. An exponential rise to a maximum value was obtained from the data, following the general equation $y = y_o + a*(1-e^{-bx})$ (Fig. 13).

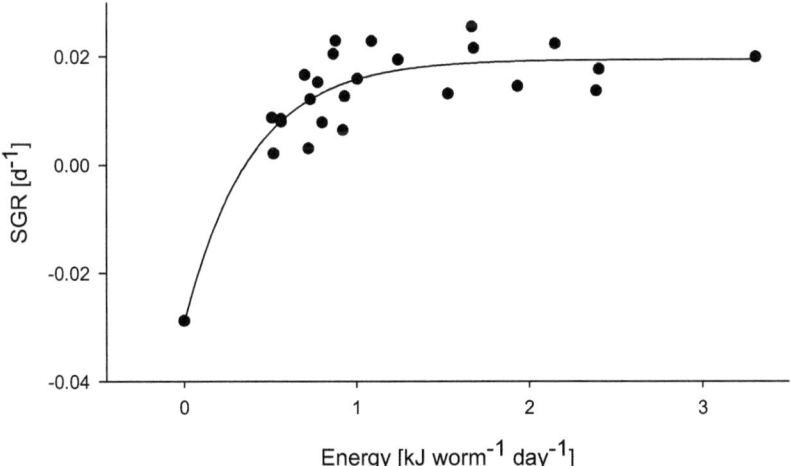

Fig. 13: Growth performance of N. diversicolor in relation to the daily supplied energy content for an individual worm. Data are taken from Bischoff (2003), regression analysis following the general equation $y = y_o + a*(1-e^{-bx})$; y_0 = -0.0291, a = 0.0486, b = 2.5774 and r^2 = 0.79

Fig. 14 presents the obtained growth curve from Bischoff (2003), combined with the specific growth rates of the worms gained from experiments within different culture systems and presented in this publication.

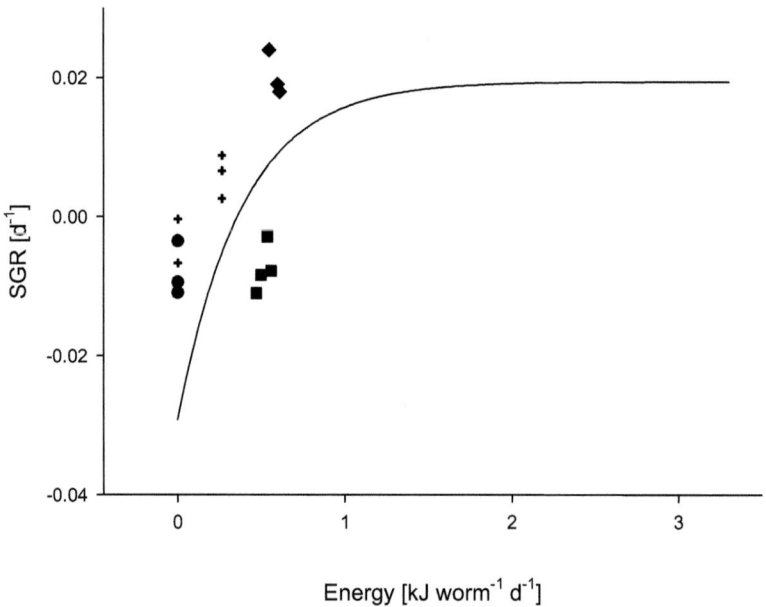

Fig. 14: Growth performance of *N. diversicolor* in relation to the daily supplied energy content for an individual worm. The graph compares the growth curve (solid line) of *N. diversicolor* according to Bischoff (2003) with results obtained during different experiments with *N. diversicolor*. Symbols represent the different experiments executed within different culture systems (● = batch culture experiment; ■ = small scale recirculating system (preference); + = medium scale recirculating system ; ♦ = MARE).

3.6 Total organic matter contents of the sediment

The total organic matter (TOM) content of the sediment obtained during the experiment within the medium scale recirculating system is given in Figs. 15 and 16 for both treatments of the experiment (presence or absence of worms in the sediment). The average TOM content of both treatments was increasing for the first 20 days from 0.4% to about 1.1%. After experimental day 20 the TOM content was stabilizing at concentrations of about 0.9% until day 36. During the last few days of the experiment the TOM content was increasing slightly to values up to 1.4% for the treatment with worms and 1.6% for the treatment without worms. No difference between treatments was detected (t-test, p = 0.509).

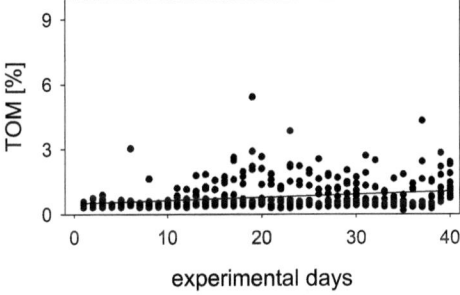

Fig. 15: Total organic matter (TOM) content obtained from the sediment of the medium scale system in the presence of worms.

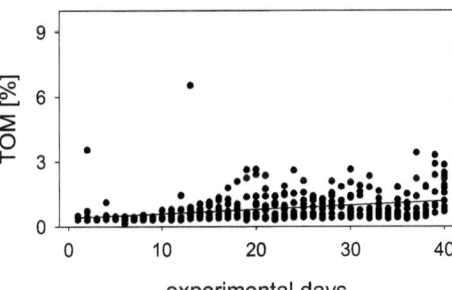

Fig. 16: TOM content obtained from the sediment of the medium scale system in the absence of worms.

Figs. 17 and 18 display the total organic matter content of the two experiments performed in the MARE-system. TOM content during the first experiment was stable for most of the experiment, with few exceptions. Overall tendency of the organic content was decreasing during the first experiment (Fig. 17). No statistically significant trend with time was detected. Also no general differences between sampling points were found (ANOVA, p = 0.954). For the second experiment of the MARE-system a statistically significant difference (ANOVA, p < 0.001; Fig. 18) was detected between sampling point 1 and all other sampling points of the bioreactor. TOM content of the first sampling point showed a statistically significant increase during the experiment. Linear regression for sampling point 1 revealed a function following the general equation $y = ax + b$ (a = 0.016, b = 2.776, r^2 = 0.12; n = 20).

Sampling points 2 to 5 showed no statistically significant trend with time (Fig. 19; t-test, p = 0.345; n = 80).

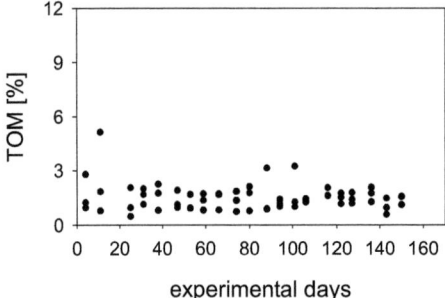

Fig. 17: TOM content of the sediment during the first experiment in the integrated multitrophic recirculating system (MARE I).

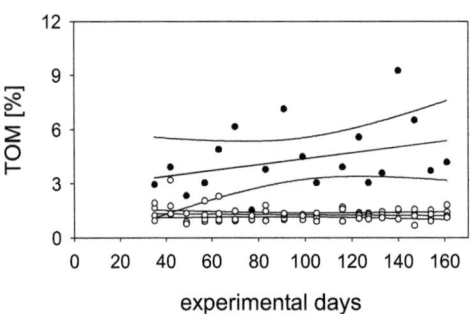

Fig. 18: TOM content of the sediment during the second experiment in the integrated multitrophic recirculating system (MARE II). (● = sampling point 1; ○ = sampling points 2 – 5). Linear regression was presented with the 95% confidence intervals.

4. Discussion
4.1 Survival of *N. diversicolor*
4.1.1 Is it possible to culture *N. diversicolor* on an exclusive diet of solid waste from fish culture?

The intention of the batch culture experiment was to investigate the growth of *N. diversicolor* fed with commercial fish feed. Due to the nitrogen content of the feed, elevated TAN concentrations occurred in the experimental tanks and caused high mortality of the worms. The control group (starvation) showed low mortalities. Abiotic conditions were similar for all tanks, implying a connection between TAN and the survival of *N. diversicolor*. This was supported by Guillen et al. (1993) and Fontenot *et al.* (1998); both studies reported about toxic effects of ammonia to marine fish and invertebrates.

It was assumed that the increase and peak of the TAN concentrations depends on the amount and type (nitrogen content) of supplied food. To achieve an improvement of the water quality, all recirculating systems used for subsequent experiments, were designed and built with a biofiltration unit for nitrification. Due to the microbial activity excreted ammonia was converted into nitrate and therefore less harmful.

The second experiment, in a small scale recirculating system, was also intended to investigate the growth of *N. diversicolor* fed with commercial fish feed and solid waste from a recirculating aquaculture system. Due to limited microbial activity at the start of the experiment, TAN concentrations were rising and thus, causing mortality of the worms. During this experiment all dissolved nutrients and the abiotic conditions were monitored regularly. Mortality increased as soon as TAN concentrations reached a certain threshold. We assumed, to avoid excess mortality the TAN concentrations should be kept as low as possible.

So far no precise statement about the toxicity of ammonia to *N. diversicolor* can be made. Further research concerning the biology of *N. diversicolor* should include investigations about the "No Observable Effects Concentration" (Ringwood and Keppler 1998), as well as the lethal concentrations, of ammonia to the worms.

The experiment within the medium scale recirculating system was designed to examine the growth of *N. diversicolor* fed with a fixed amount of energy, supplied in the form of solid waste. Abiotic parameters and dissolved inorganic nutrient concentrations were suitable for the worms and consequently, positive worm growth could be observed.

We assumed that mortality in the beginning of all experiments was caused by handling the worms prior to the experiments. After a period of approx. 5 days this mortality disappeared. This assumption is supported by the low TAN concentrations recorded during the initial phase of each experiment.

4.1.2 Which impact has the type of sediment on the survival of *N. diversicolor*?

An experiment to investigate the impact caused by the substrate on the survival of *N. diversicolor* obtained on average survival rates between 60 and 90% depending on the applied substrate. No substrate at all showed lowest survival rates. Results reported by Batista and co-authors (2003), presented average survival rates for fed worms without sediment of approx. 35%. During daily monitoring and sampling of the experimental tanks, not all carcasses could be detected assuming cannibalism within these tanks. Cannibalism for *N. diversicolor* was also reported by Hartmann-Schröder (1996) and Batista *et al.* (2003). It was concluded that cannibalism is fostered in the absence of a suitable substrate.

Highest survival rates were obtained for coarse sand (grain size 2 – 4 mm) and Aquaclay®, which both showed average survival rates around 90%. This could be explained by the good water exchange within these substrates and thus, constant low TAN concentrations surrounding the worms. For fine sand survival rates around 80% were recorded, implying a lower potential for this type of sediment compared to coarse sand and Aquaclay®. Nevertheless, after comparing the combined performance of all substrates (survival and growth), fine sand was chosen for further experiments. (For the substrates impacts on growth see section 4.2.1).

With increasing system volume more stable conditions within the system, and in particular within the sediment, evolved leading to decreased mortality rates.

4.2 Growth of *N. diversicolor*

4.2.1 What influence has the type of sediment on the growth of *N. diversicolor*?

The type of sediment seems to influence the growth of *N. diversicolor*. Therefore, during the experiment in the small scale recirculating system with different substrates, the growth of the worms was also examined.

Growth performance in all substrates was poor, showing negative growth rates for all tanks. However, a tendency revealing the potential of the substrates was noticeable. Fine sand showed the lowest weight loss, followed by no substrate at all and ceramic filter media. The highest weight loss occurred for the coarse sand. It seemed that the worms utilized more energy for moving and stabilizing this type of sediment as they built their tubes. Burrows of *N. diversicolor* are lined on the inside with mucus for stabilisation (Hartmann-Schröder 1996). This energy requirement for mucus production seems to increase with increasing grain size of the substrate. The worms of the treatment group without substrate at all, may have to invest energy to stabilize the supplied solid waste to form a substrate, suitable for their required conditions.

Personal observations of substrate-balls supported this assumption. This energy was also lost for the metabolism of the worms.

Besides the metabolic costs of the worms, the lack of essential nutritional elements may explain the poor growth. All small scale experiments were using sediments with little microbial activity at the start of the experiments, whereas the experiment within the medium scale recirculating system and the two MARE experiments used sediments which had an active microbiology, which means the microbial community was allowed to develop for a period of several weeks. Comparing the results from the small scale recirculating system (preference) and the second MARE experiment showed that although supplying approximately the same amount of energy to the worms, different growth rates were recorded. This implies that besides supplied energy and nutrients, which were adequate to meet the requirements of the worms, some other essential elements were missing in clean sediments, which need to be delivered by some other forms of food e.g. bacteria or protozoa. It is recommended for the culture of *N. diversicolor*, fed with solid waste from aquaculture as the exclusive food source, to allow the sediment to develop an active microbiology and thus, a gardening of meiobenthic organisms by *N. diversicolor* (Olivier et al. 1995) (see also section 4.2.3). Experiments which run longer than 20 days showed better growth compared to short term experiments. This is in agreement with the assumption for the necessity of a well established microbial loop within the sediment. We assumed that both, the type and the age of the sediment effect the growth of the worms.

4.2.2 What is the optimum achievable growth under the applied conditions?

Growth of *N. diversicolor* within the diverse culture systems was quite different, ranging from negative growth rates up to 0.025 d^{-1}, calculated with the arithmetic mean weight from the second experiment of the MARE system. The positive specific growth rates obtained under cultural conditions were comparable to those cultured under similar conditions with other food sources (Nielsen et al. 1995; Olivier et al. 1995; Saiz-Salinas and Frances-Zubillaga 1997). Specific growth rates were higher compared to natural populations which range between 0.008 and 0.015 d^{-1} (Chambers and Milne 1975; Vedel and Riisgard 1993). Fig. 15 compares all obtained specific growth rates with a growth curve of *N. diversicolor* under similar conditions taken from Bischoff (2003). During the designing process of the experiments we assumed that supplied energy exceeding a daily amount of 1 kJ per individual worm was not desirable because no increase in growth could be achieved by the additional energy. Therefore all obtained growth rates are located below 1 kJ of supplied energy. Most of the presented specific growth rates were higher than expected from the calculated growth curve, except the results of the experiments from the small scale recirculating system used for the evaluation of the preferred sediment. This proves the overall fit of the equation describing the growth of *N. diversicolor* with respect to the supplied energy.

Applying the observed specific growth rates leads to the assumption, that worms will take around 100 days to reach initial reproducers size. This is significantly faster compared to natural occurring conditions, where longevity of the worms ranges between 12 and 36 months, depending on the geographical locality (Dales 1950; Chambers and Milne 1975; Mettman et al. 1982).

4.2.3 Are the applied conditions adequate to complete a lifecycle of *N. diversicolor*?

The growth results obtained from the integrated system during the second MARE experiment prove that *N. diversicolor* was able to grow until sexual maturity and consequently, complete a lifecycle under the applied conditions. Assuming high specific growth rates of 0.05 to 0.06 d^{-1} (Riisgard et al. 1996; Costa et al. 2000) *N. diversicolor* will take from hatch until reproduction approx. 75 to 80 days. Weight at maturity was assumed to be around 750 mg. The completion of a lifecycle observed under the applied conditions during the MARE experiment took approx. 110 days (estimated from the initial average weight of subsequent worm generations within the *Nereis* tank). This was longer than estimated for optimal conditions. There might be two possible explanations for this elongated time. One explanation could be that some minor elements necessary for a good growth performance are limited but not missing, otherwise the life cycle could not be finished at all. Possible compounds, to explain the prolonged lifecycle, might be micronutrients such as vitamins or trace elements, e.g. sulphur, copper, manganese or chromium. Trace elements accomplish different important physiological processes within the body and under limiting conditions they interact with different metabolic processes such as limited growth (sulphur), lack of different enzymes (copper or manganese) or disturbed energy metabolism (chromium) (www.vitfit.de/spurenelemente.htm). The second explanation could be an accumulation of heavy metals by *N. diversicolor*. This was demonstrated for cadmium, chromium, copper, lead, mercury, nickel and zinc (Kolbe 1995; Ferns and Anderson 1997; Baeyens *et al.* 1998; Hursthouse *et al.* 2003). Bordin *et al.* (1994) showed that in vitro intoxication with Cd, Cu and Zn strongly affects the original metal intracellular partitioning between the insoluble phase, the high molecular weight proteins and the heat-stable cytosol. The neuron structure of *Viviparus ater* appeared particularly affected by lead exposure as reported by Fantin and Franchini (1990), which led to ultrastructural changes in the ganglia of this aquatic snail. Hateley *et al.* (1989) described the increase in tolerance of *N. diversicolor* to different substances such as Cu and Zn. This is in agreement with Bryan (1975) who further suggested a possible tolerance to the toxic effects of silver and arsenic. Additionally, Septier *et al.* (1991) proposed that the gut of *N. diversicolor* could be a detoxification and excretory site. This could explain the time laps until reproduction caused by adaptation processes of the worms.

Fig. 13 presents besides the average bodyweight of the worms the size classes obtained during the experiments within the MARE-system. At the start of the second experiment, a high percentage of small worms in the smallest size class were detected. These worms grew during the experiment and shifted to larger size classes. After reproduction large individuals of *N. diversicolor* died and the offspring was first recorded 20 days after the assumed spawning event at the following biomass determination of the worms. The supplied amount of energy and nutrients delivered by the solid waste from the fish tanks in combination with trace elements and vitamins from bacterial loops within the sediment seemed sufficient to allow the observed positive growth rates of *N. diversicolor* during this trial.

Factors controlling the reproduction of *N. diversicolor* are not fully understood. Future research should investigate the reproduction of *N. diversicolor* with focus on external factors controlling reproduction.

4.2.4 Is the total organic matter content of the sediment a reliable indicator for the consumption of solid waste by *N. diversicolor*?

The total organic matter (TOM) content of the sediment analysed during the experiments was used as an indicator for the consumption of applied foods by the polychaetes. TOM contents was in most of the experimental cases low, but occasionally high concentrations comparable to natural or even polluted environments were observed. During the experiment with or without the presence of worms, the total organic content of the sediment showed no detectable differences. Concentrations of TOM within the sediment were low compared to natural conditions since *N. diversicolor* favours higher organic contents as described by Hansen and Kristensen (1997). Positive worm growth and a biomass increase could be recorded. An assumed influence on the organic matter content of the sediment, caused by the consumption of the worms should have been noticeable during this experiment. Results did not show such an influence.

During the investigations within the multitrophic integrated recirculating system two scenarios concerning the TOM content of the sediment were detected. Throughout the first experiment, the stocked fish biomass was low and thus, the amount of solid waste originating from the fish transferred to the worm bioreactor was also low. As a consequence the TOM content of the sediment within this bioreactor kept constant. For more details see Wecker *et al.* (2006) in Kube (2006). Through the growth of the fish within the multitrophic integrated system, the fish biomass increased and so did the amount of solid waste, transferred to the worm tank. TOM content beneath the water inlet of the worm bioreactor (sampling point 1) increased significantly during the second experiment in the integrated system, whereas the rest of the bioreactors sediment did not change. The explanation therefore could be the settling velocities of the solid waste. Most of the particles settle in the vicinity of the water inlet and as a result, increasing the total organic matter content in that area. Supporting to the increase of organic matter in the first section of the bioreactor could be the reproductive behaviour of *N. diversicolor*. Reproduction is monotelic and after reproduction mature animals die. As soon as ragworms begin to reach maturity, hormonal changes cause their bodies to alter. The digestive system of female worms breaks down, to enable large numbers of eggs to be produced (Hartmann-Schröder 1996; UK Marine 2006). Therefore female worms stop feeding as soon as egg production commenced. According to Dales (1950) populations possessing more females compared to males can be observed. Consequently, more than half of the solid waste formerly consumed by worms is allowed to accumulate within the sediment. Due to the increasing TOM content hypoxic conditions or even anoxic conditions could develop in the sediment but these conditions could not be detected in the overlaying water column. It is assumed that microbial processes within the sediment such as denitrification occurred. Conditions favouring denitrification are described by Rheinheimer et al. (1988) and are the availability of organic matter within the sediment, the presence of nitrate and low oxygen contents. All three conditions were fulfilled and therefore denitrification was expected to occur. The decrease of nitrate during the second MARE experiment (see Chapter 2) supported the assumption that denitrification within the sediment occurred. Denitrification was fuelled by the organic matter within the sediment and thereby degrading the organic matter.

Therefore, the TOM content in the sediment can not be used as a reliable indicator for the consumption of solid waste by *N. diversicolor*. The comparison of fatty acid compositions derived from fish feed, solid waste, sediments and worm tissue, as described by Bischoff et al. (see Chapter 4) seems to be a more reliable indicator for the consumption of solid waste by rag worms.

5. Conclusions

By the combination of several experiments performed in different culture systems a series of questions concerning the culture of *N. diversicolor* with solid waste as food source could be addressed:

The type and amount of feed applied to the system controls the concentrations of dissolved inorganic nutrients. Ammonia was assumed to be the most crucial dissolved nutrient, affecting the survival of *N. diversicolor*. Ammonia concentrations resulting from pelleted fish feed surmounting concentrations resulting by nutrient leaching from excreted solids. To avoid excess mortality rates, it is recommended to monitor ammonia concentrations thoroughly and keep them at low levels. For other monitored dissolved nutrients, no such effect could be observed.

The type of sediment used for worm culture affects their survival and growth. Fine grained sediment represents the best choice for the culture of *N. diversicolor*.

Although anoxic microniches might appear in this type of sediment, the metabolic costs for living and moving within this type of sediment are the lowest and polychaetes are able to tolerate severe conditions such as low oxygen concentrations. Cannibalism might be a problem but can be avoided by an appropriate sediment and sufficient food supply. The conditions seem to be adequate to allow growth of the worms which is higher compared to natural conditions. A completion of the life cycle under the applied conditions could be achieved in less than four months. This is much faster than under natural conditions, where the longevity of the worms ranges between 12 and 36 months depending on the geographical locality.

Concluding, *N. diversicolor* is an appropriate organism for integrated aquaculture and its culture is one possibility to reduce the amount of solid waste originating from fish and/or crustacean culture within recirculating systems.

Chapter 4

Effects of different diets on the fatty acid composition of *Nereis diversicolor* (O. F. Mueller, 1776) with possible implications for aquaculture

Bischoff A.A., Fink P. and Waller U.

Abstract

The potential of solid waste, originating from a fish culture system, i.e. faecal material, uneaten food pellets and bacterial biofilms was examined as food source for *Nereis diversicolor*, which could be a valuable food for fish providing a suitable fatty acid signature. Therefore, the fatty acid profiles from organisms and materials involved in the process were examined. Different sets of samples have been analysed including the pelleted fish feed, fish liver, faecal material and several batches of worm samples collected from the wild and an integrated recirculating aquaculture system.

Numerous fatty acids were detected including saturated (C16:0 and C18:0), monounsaturated (C18:1 (n-7) and C18:1 (n-9/n-12)) and essential polyunsaturated (C18:3 (n-3), C20:4 (n-6), C20:5 (n-3) and C22:6 (n-3)) fatty acids.

A recycling of valuable feed nutrients such as fatty acids can be achieved in integrated aquaculture and thus, a production of raw materials demanded for animal nutrition in integrated aquaculture is possible.

1. Introduction

Aquatic food webs are complex structures, which are combined of organisms belonging to different trophic levels (Belgrano 2005). In natural food webs, most of the nutrients are retained, because they are a resource for other members of the food web. The technical application of a simple food web structure in aquaculture systems is named here an integrated aquaculture system (Neori *et al.* 2004; Schneider *et al.* 2005) which is gaining importance, as conventional aquaculture is still a potential thread to natural ecosystems (Edwards and Pullin 1990; Costa-Pierce 1996; Gyllenhammar and Håkanson 2005). Besides the environmental aspect, the variety of organisms which can be maintained in integrated systems also offers the advantage that valuable industrial raw materials could be produced in addition to biomass. A prominent raw material which can be produced in integrated aquaculture systems are fatty acids which are demanded in human and animal nutrition.

During the past decades, lipids and especially fatty acids came into focus as they are of great importance for the food web. Fatty acids can be used as biological markers and diet indicators in marine ecosystems (Sargent *et al.* 1987). Fatty acids from marine organisms contain 14 to 24 carbon atoms and have varying degrees of unsaturation (Budge *et al.* 2002). Marine animals have limited abilities of fatty acid synthesis (elongation, desaturation). Marine fish for example need to retrieve fatty acids from lower trophic levels; fatty acids are incorporated with little or no modification. Thus, fatty acids are useful indicators for the origin of food and dietary fatty acid combinations are characteristic fatty acid signatures (Iverson 1993). Numerous studies have demonstrated that fatty acid signatures can be passed from prey to predator (Fraser et al. 1989; Graeve et al. 1994; Kirsch et al. 1998; Kirsch et al. 2000). Once a fatty acid signature was established for prey items in a food web it can be used to trace its pathway (Budge et al. 2002).

The marine polychaete N. diversicolor which is rich in fatty acids (Luis and Passos 1995; Olive 1999) inhabits the temperate coast of the North Atlantic. It lives in tubes in sandy mud, gravel and clay. The polychaete is able to change the feeding mode from filter feeding to deposit feeding or active carnivorous feeding on small meiobenthic organisms, depending on the environmental conditions (Hartmann-Schroeder 1996). Due to the way of life of N. diversicolor it seemed to be a possible organism to produce fatty acids from fish waste collected from a primary fish aquaculture system. This way valuable food items can be produced from recycling materials. Marine polychaets are essential food for a variety of cultured species like marine prawns or flatfish (Pousao et al. 1995; Sudaryono et al. 1995; Costa et al. 2000; Wouters et al. 2002).

Luis and Passos (1995) showed that the composition of diets is a decisive factor determining fatty acid composition of N. diversicolor. Their results confirmed the detritivorous feeding mode as well as the sediment swallowing of N. diversicolor, which was described by Mettam (1979) and Fauchald and Jumars (1979) for different worm populations. In this study, the potential of solid waste, originating from a fish culture system, i.e. faecal material, uneaten food pellets and bacterial biofilms was examined as food source for N. diversicolor which could be a valuable food for fish providing a suitable fatty acid signature. The proof was made by the examination of fatty acid profiles from organisms and materials involved in the process.

2. Materials and Methods

N. diversicolor were sampled along the French Atlantic coast and transported alive into the lab facilities. One batch of animals (batch (N)) got immediately quick-frozen and was stored at − 40 °C until sample preparation and analysis. The worms were transferred into a research-scale integrated recirculating system combined of two fish tanks, a sedimentation tank, a foam fractionator, an aerobic biofilter, and an algae tank (*Ulva lactuca* or *Solieria chordalis*). The recirculation system was stocked with sea bream, *Sparus auratus*, which were grown from 4.5 ± 0.8 g to 607 ± 91 g living weight. Commercial fish feed Biomar *Ecostart 17* (pellet size 1.1 to 2 mm) and *Aqualife 17* (pellet size 3 to 6 mm) was administered every day to the fish. The worms were maintained in the rectangular sedimentation tank. Baffle plates forced the water to flow in a sinuous line so that floating particles (faecal material, fish waste) were sinking down onto the sediment. The sediment provided living space for the worms in which they buried tunnels. During the course of the experiment solid waste was the exclusive food source. For further description of the integrated recirculating system we refer to Chapter 2.

Different sets of samples have been analysed including the pelleted fish feed, fish liver and faecal material. Fish liver samples were taken from euthanized *S. auratus* (555 ± 106 g). Faecal material was collected after dissection from the last few centimetres of the hindgut. Further samples were taken from the sediment in the sedimentation tank. The first set of worm samples from wild caught individuals were taken immediately after the arrival at the lab facilities and will be referred to as batch (N). Further sets of samples were taken at four discrete samplings times by siphoning out the worms from the sediment. Samples were taken in June 2004 (A) ten days after transfer into the experimental RAS and in December 2005 (B). Final samples were taken in the beginning (C) and at the end of February 2006 (D). In total four generations were maintained in the RAS throughout the experimental period. The animals of samples B, C, and D were offspring from the wild caught population from the French Atlantic coast. Table 1 characterises the conditions present prior to the collection of the worms. The animals were analysed as whole animals.

Table 1: Characterisation of worm samples, environmental and food conditions in the RAS (Recirculated Aquaculture System) during the course of experiments.

Worm sample	Abiotic conditions	food conditions	Sample size
Batch (N)	French Atlantic coast	natural food	5 Individuals
Batch (A)	S 29.5 ± 0.5 T 19.4 ± 0.3 pH 8.2 ± 0.1	solid waste from a RAS (10 days)	6 Individuals
Batch (B) (C) (D)	S 25.0 ± 1.0 T 18.8 ± 0.4 pH 7.9 ± 0.2	solid waste from a RAS (self sustaining population > 18 months)	5 Individuals 6 Individuals 30 Individuals

Table 2: List of fatty acids that had been quantified after extraction by gas chromatography.

Systematic name	Scientific name	Common name
Saturates		
C14:0	Tetradecanoic acid	Myristic acid
C15:0	Pentadecanoic acid	
C16:0	Hexadecanoic acid	Palmitic acid
C18:0	Octadecanoic acid	Stearic acid
C20:0	Eicosanoic acid	Arachidic acid
Monounsaturates		
C16:1	9-Hexadecenoic acid	Palmitoleic acid
C17:1	8-Heptadecenoic acid	
C18:1(n-9/n-12)	9-Octadenoic acid	Oleic acid
C18:1(n-7)	11-Octadecenoic	Vaccenic acid
C20:1(n-9)	11-Eicosenoic acid	
C22:1(n-9)	13-Docosenoic acid	Erucic acid
Polyunsaturates		
C18:2(n-6)	9,12-Octadecadienoic acid	Linoleic acid
C18:3(n-3)	9,12,15-Octadecatrienoic acid	α-Linolenic acid
C20:3(n-6)	8,11,14-Eicosatrienoic acid	Homo-γ-linolenic acid
C20:4(n-6)	5,8,11,14-Eicosatetraenoic acid	Arachidonic acid (AA)
C20:5(n-3)	5,8,11,14,17-Eicosapentaenoic acid	Eicosapentaenoic acid
C22:2(n-6)	13,16-Docosadienoic acid	Docosadienoic acid
C22:6(n-3)	4,7,10,13,16,19-Docosahexaenoic acid	Docosahexaenoic acid

Fatty acids (Table 2) were extracted with dichloromethane:methanol (2:1, vol/vol) as described by von Elert & Stampfl (2000). Polyunsaturated fatty acids (PUFAs) were quantified as fatty acid methyl esters (FAMEs) using a HP 5890 Series II GC (Agilent Technologies, Waldbronn, Germany) equipped with a DB-225 fused silica column (J&W Scientific, Folsom, USA) and a flame ionisation detector with heptadecanoic acid methyl ester and tricosanoic acid methyl ester as internal standards (von Elert and Stampfl 2000). Identification of the FAMEs was based on comparison of retention times to those of reference compounds. The amount of individual fatty acids is given as share (percentage) of the total amount of all measured fatty acids.

3. Results

The major saturated fatty acid was C16:0, palmitic acid which may be a precursor of longer and unsaturated fatty acids. The content averaged around 26% of the fatty acid composition in all samples taken from fish, sediments, and *N. diversicolor* (Tab. 3). The highest amount was found in fish faeces (34%); comparably low amounts were determined in the liver tissue of the fish (22%). The palmitic acid content in wild *N. diversicolor* was comparable. At the end of the experimental trials in the integrated RAS, the fatty acid content of the saturated palmitic acid under cultural conditions (24%) was slightly lower compared to natural conditions (26%). A considerable amount of palmitic acid was available in the sediment and as mentioned before the highest amount was determined in fish faeces.

The group of monounsaturated fatty acids was dominated by the C18:1 group. The average content of C18:1 (n-7) and C18:1 (n-9/n-12) was 5 and 10% in all samples including the sediment and fish faeces. Vaccenic acid (C18:1 (n-7)) was higher in wild *N. diversicolor* (batch (N) (9%)) and *N. diversicolor* which were cultured for short time in the RAS (batch (A)) (8%) compared to all other samples (5%). While fish (liver) contained 16% oleic acid (C18:1 (n-9/n-12)), the wild *N. diversicolor* (batch (N)), and the individuals that had been maintained for short time in the RAS (batch (A)) had much lower contents (4%) in their tissue. Towards the end of the experiment in the integrated RAS the oleic acid content in *N. diversicolor* increased to values between 9 and 14% (batch (C) and (D)).

Alpha-Linolenic acid (C18:3 (n-3)) was in the range of 2 – 3% in most of the samples. An extremely high proportion was found in the sediment (22%) which is tenfold higher compared to all other samples. Alpha-Linolenic acid is an important precursor of longer chained polyunsaturated fatty acids in fish and may not be excreted via faeces. It could be a metabolic product of micro organisms in the sediment.

Eicosapentaenoic acid (EPA) (C20:5 (n-3)) which is essential in many fish was detectable in higher proportions in all *N. diversicolor* samples but in lower proportions in fish liver and fish feed. The content in *N. diversicolor* (39%) was almost four times higher than in fish liver (11%) and only little EPA was found in fish waste and sediments which is expected from an essential nutrient.

Tab. 3: Fatty acid composition (% ± SD) of fish feed, fish (*S. aurata*) liver, fish faeces, sediment, and *N. diversicolor* (from the field, batch (N) and (A), and an integrated aquaculture systems batch (B) to (D)).

Sample	Fatty acid					
	Saturated	Monounsaturated		Polyunsaturated		
	C16:0	C18:1 (n-7)	C18:1 (n-9/n-12)	C18:3 (n-3)	C20:5 (n-3)	C22:6 (n-3)
Fish						
feed	25.6 ± 1.1	3.0 ± 0.3	11.1 ± 4.5	1.9 ± 1.1	11.9 ± 1.3	13.5 ± 1.4
liver	21.5 ± 0.9	3.0 ± 0.2	15.8 ± 1.6	2.4 ± 0.4	11.0 ± 0.7	20.1 ± 2.0
faeces	34.3 ± 4.4	2.9 ± 0.3	11.1 ± 2.6	2.1 ± 1.6	5.1 ± 1.2	12.4 ± 6.3
Sediment	28.7 ± 2.3	6.7 ± 1.0	10.0 ± 2.4	22.2 ± 2.4	1.8 ± 0.6	0
Nereis diversicolor						
batch N	26.0 ± 0.7	8.6 ± 0.3	3.9 ± 0.3	0	38.7 ± 1.4	0
batch A	26.2 ± 1.3	7.9 ± 3.9	4.4 ± 0.3	0	39.0 ± 2.5	0
batch B	23.2 ± 0.9	5.0 ± 0.4	13.8 ± 3.0	2.2 ± 0.5	25.6 ± 1.7	5.8 ± 1.8
batch C	22.7 ± 0.8	5.5 ± 0.4	9.7 ± 1.9	3.3 ± 1.2	21.1 ± 1.5	5.0 ± 0.5
batch D	25.6 ± 1.7	5.1 ± 0.9	8.7 ± 1.4	2.6 ± 1.0	24.9 ± 2.4	4.8 ± 0.8

Docosahexaenoic acid (DHA) (C22:6 (n-3)) which is also essential for many fish was detectable in smaller proportions in cultured *N. diversicolor* (batch (B) to (D)) but in higher proportions in fish liver, faeces and fish feed. The content in cultured *N. diversicolor* (5%) was four times lower than in fish liver (20%). No DHA was found in sediments which was expected from an essential fatty acid, but also in wild caught *N. diversicolor* no DHA was detectable.

Detailed fatty acid profiles for the pelleted feed and fish are shown in Figs. 1 and 2. Fatty acids are digested and catabolized with great ease in fish and so they serve as metabolic energy source. The administered feed had a high amount of C16:0, palmitic acid (25.6 ± 1.1%). The average amount of stearic acid (C18:0), the monounsaturated C16:1, palmitoleic acid and C18:1 (n-9/n-12), oleic acid were 9.3 ± 1.6, 5.9 ± 0.3 and 11.1 ± 4.5%, respectively.

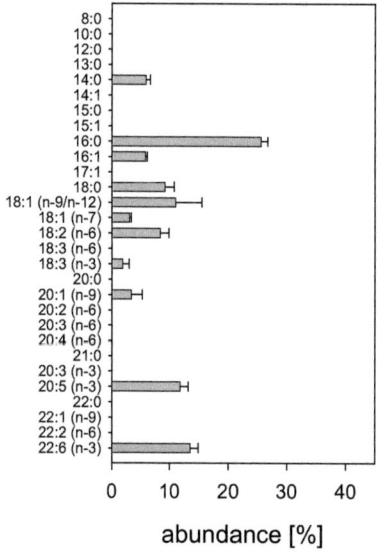

Fig. 1: Fatty acid profile of the pelleted fish feed Biomar *Aqualife* 17 applied for the culture of *S. aurata*. Bars represent means + SD. (n = 6)

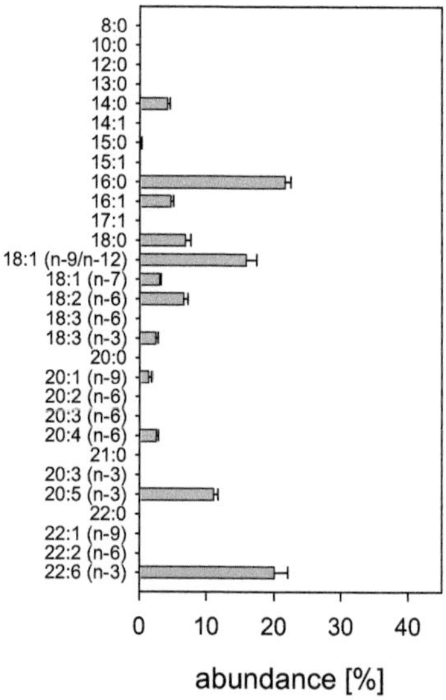

Fig. 2: Fatty acid profile of fish liver (*S. aurata*) cultured with Biomar *Aqualife* 17. Bars represent means + SD. (n = 14)

The pelleted feed contained 11.9 ± 1.3% EPA and 13.5 ± 1.4% DHA which are likely to be essential in marine fish. The potential precursor of these two fatty acids is α-linolenic acid which represented 1.9 ± 1.1% of total fatty acids. In the fish liver (Fig. 2) a marked increase in DHA was determined (20.1 ± 2.0%) which may be explained by synthesis via the C18:3 (n-3) – C20:4 (n-6) – C20:5 (n-3) – C22:6 (n-3) pathway. Precursor and intermediate products were determined in the range of 2.4 ± 0.4, 2.5 ± 0.4 and 11.0 ± 0.7%, respectively.

The fish faeces (Fig. 3) contained a high amount of fatty acids. Besides the dominating saturated palmitic acid (C16:0) also C14:0, myristic acid and C18:0, stearic acid were found at 6.9 ± 1.9% and 10.9 ± 1.3%. It was conspicuous that the fish faeces contained 34.3 ± 5.5% palmitic acid while the amount of stearic, palmitoleic, and oleic acid remained in the range determined in the pelleted feed.

Thus, the increase of palmitic acid cannot be explained by digestive or absorptive processes.

The main monounsaturated fatty acid in fish faeces was C18:1 (n-9/n-12), oleic acid (11.1 ± 2.6%). EPA and DHA were determined to 5.1 ± 1.2 and 12.4 ± 6.3%, respectively. It is obvious that significant amounts of feed grade fatty acids were not digested and metabolized and are usually discarded from RAS systems.

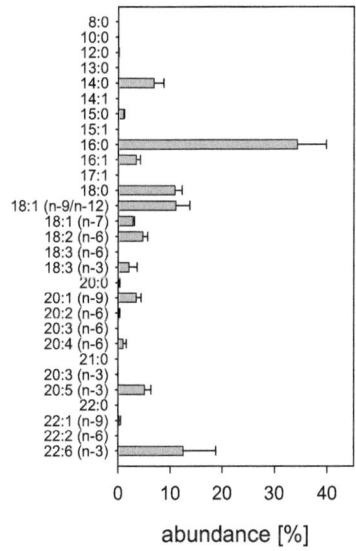

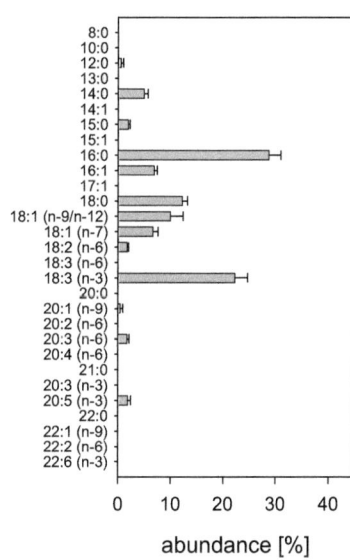

Fig. 3: Fatty acid profile of fish faeces extracted from *S. aurata*. Bars represent means + SD. (n = 9)

Fig. 4: Fatty acid profile of the sediment collected from the sedimentation tank of the integrated RAS. Bars represent means + SD. (n = 5)

In the organic fraction of the sediment within the sedimentation tank of the RAS C16:0, palmitic acid (28.7 ± 2.3%) and C18:3 (n-3), α-linolenic acid were dominating (22.2 ± 2.4%, Fig. 4). Smaller amounts of C14:0, myristic acid (5.0 ± 0.8%), C16:1, palmitoleic acid (6.9 ± 0.6%), C18:0, stearic acid (12.2 ± 1.0%), and C18:1 (n-7), vaccenic acid (6.7 ± 1.0%) were determined. The great fraction of C18:3 (n-3), α - linolenic acid is striking, since only small amounts of C18:3 (n-3) could be derived from the faeces of the fish; as C18:3 (n-3), α - linolenic acid represented only 2.1 ± 1.6% in the fish faeces (Fig. 3). The same applies to C18:1 (n-7), vaccenic acid which is increasing from 2.9 ± 0.3% in fish faeces (Fig. 3) to 6.7 ± 1.0% in the sediment of the RAS. Another striking finding was that DHA was missing in the sediment profiles although they had been available from the fish faeces (Fig. 3, 12.4 ± 6.3%).

The fatty acid profile of *N. diversicolor* from natural environments is shown in Fig. 5. The fatty acid composition had hardly changed after the animals had been maintained for short time (10 d) in the RAS (Fig. 6). In all samples C16:0, palmitic acid ranged around 26% (26.0 ± 0.7%, 26.2 ± 1.3%, Fig. 5 and 6). The fraction of EPA was appox. 39% (38.7 ± 1.4%, 39.0 ± 2.5%, Fig. 5 and 6) which was high compared to all other *N. diversicolor* samples (Fig. 7).

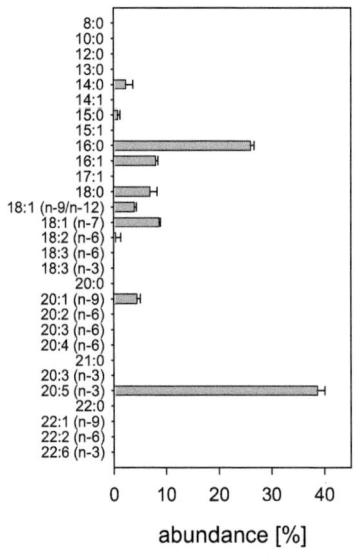 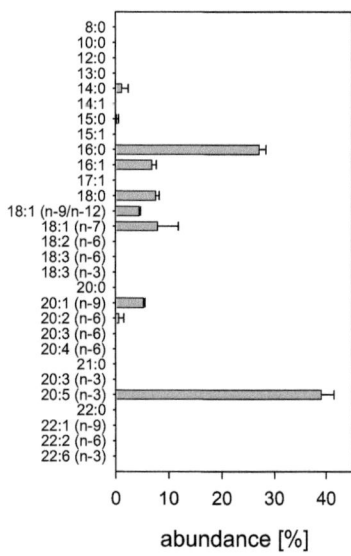

Fig. 5: Fatty acid profile of *N. diversicolor* under natural conditions (batch (N)). Bars represent means + SD. (n = 5)

Fig. 6: Fatty acid profile of *N. diversicolor* after a short term (10 days) in an RAS (batch (A)). Bars represent means + SD. (n = 6)

The fatty acid profiles of *N. diversicolor* that had been maintained in the experimental integrated RAS showed considerable changes.

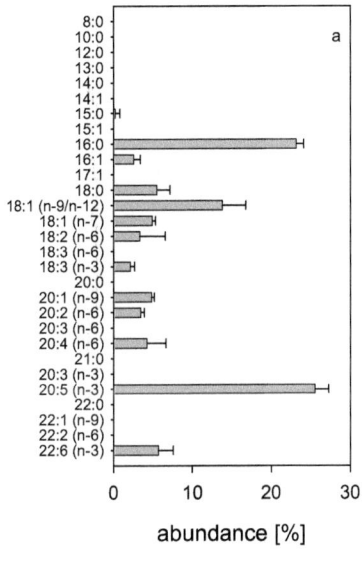

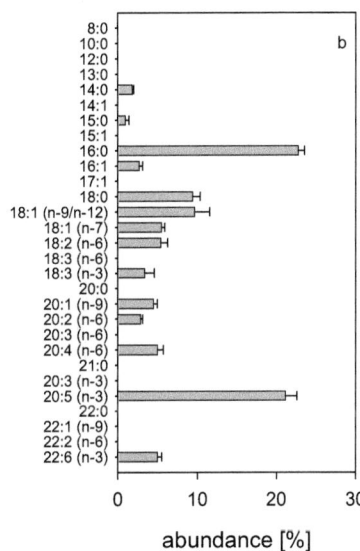

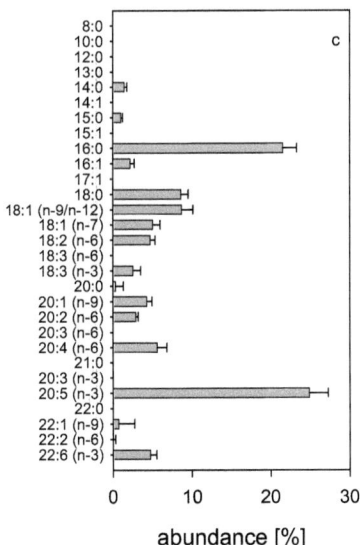

Fig. 7: Fatty acid profile of *N. diversicolor* cultured in the integrated RAS. a) fatty acid profile of batch (B), collected in the mid of December 2005 (n = 5); b) fatty acid profile of batch (C), worms were collected in the mid of February 2006 (n = 6) and c) fatty acid profile of batch (D), collected in the end of February 2006. (n = 30). Bars represent means + SD.

In all experimental groups from the MARE experiment *N. diversicolor* contained the full set of essential fatty acids including the C22:6 (n-3), docosahexaenoic acid (4.9 ±

1.0%, Fig. 7) which had not been available from the organic matter in the sediment (Fig. 4).

Precursor (C18:1 (n-9/n-12), oleic acid, 9.5 ± 2.3%; C18:3 (n-3), α - linolenic acid, 2.7 ± 1.0%) and intermediate products (C20:4 (n-6), arachidonic acid, 5.4 ± 1.4%) were existing in worms so that a synthesis of DHA appears to be likely.

Fig. 8 displays the absolute amounts of fatty acids for all analysed sampling materials. All values are displayed as means ± standard errors in µg fatty acids per mg dry weight. Fish feed delivers an amount of 24.5 ± 0.2 µg FA / mg dry weight to the fish. For the fish muscle 19.7 ± 0.5 µg FA / mg dry weight and for the fish liver 46.9 ± 2.1 µg FA / mg dry weight were detected. Fish faeces samples revealed amounts of 39.6 ± 5.3 µg FA / mg dry weight, whereas for the sediment samples amounts of 4.1 ± 0.3 µg FA / mg dry weight were obtained. Significant differences (t-test, p < 0.001) were detected between wild caught (16.6 ± 6.9 µg FA / mg dry weight; n = 11) and cultured (27.1 ± 1.0 µg FA / mg dry weight; n = 42) *N. diversicolor*.

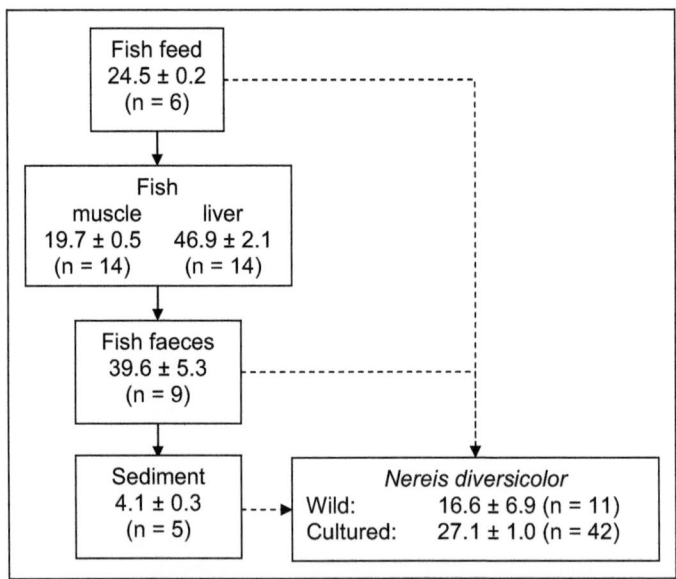

Fig. 8: Absolute amounts (mean ± SE) of fatty acids in [µg FA / mg dry weight], detected during the analyses of all sampling materials. Solid lines indicate the assumed pathway of the fatty acids with respect to the fish and the dashed lines indicate the assumed nutritional sources of *N. diversicolor*.

4. Discussion

Sea bream need the supply of essential unsaturated fatty acids (Caballero et al. 2004; Ibeas et al. 2000) which are supplied by the feed. Despite the fact that unsaturated fatty acids are essential for the properties of biomembranes which mediates physiological processes, C 20-fatty acids of the omega-6 group are precursors for anti-inflammatory prostaglandins (Gill and Valvivety 1997, Lehninger et al. 1993). At last, an unbalanced supply with fatty acid may cause metabolic disorders (Caballero et al. 2004) which may also diminish the value of the culture product. In this study the feed pellets contained 2.5 ± 0.5 % of dry matter fatty acids or 0.5 % of unsaturated fatty acids, which were taken up. However, the significant amount of fatty acids in the faeces indicates that fish had been fed in excess which is unnecessary and not acceptable in view to the natural resources. It is needed to investigate the true metabolic fatty acid demand of cultured fish in order to minimise the losses from intensive aquaculture operations. There are indications on the other side that an overfeeding of essential fatty acid can have negative effects on growth in S. aurata (Ibeas et al. 2000) and may cause severe organ alterations under winter conditions (Ibarz et al. 2005).

The projected increasing demand for fishmeal and fish oil for aquaculture feeds has led to the concern that the farming of carnivorous and omnivorous fish will place pressure on the wild fish stocks because feed grade fishmeal and fish oil are raw material for aquafeeds. In reverse, the lack of fishmeal and fish oil derived from wild fish may constrain future aquaculture development.

It was found in this study that fish faeces from aquaculture with carnivorous fish contained considerable amounts of essential fatty acids. Therefore, the reuse of wastes from such aquaculture will become an important factor in order to recycle valuable raw organic matter. Fish waste is a commodity which can be recycled thereby contributing in many cases to sustainable aquaculture development.

Integrated aquaculture has gained interest during the past decade. The focus was mainly on the removal of dissolved nutrients such as ammonia, nitrate and phosphate (Edwards 1998; Troell et al. 2003; Chopin and Bastarache 2004; Buschmann et al. 2005; Matos et al. 2006) from effluent water; particulate matter such as faecal material and uneaten food gained less attention (Erler et al. 2000; Batista et al. 2003). However, particular waste from fish production contains valuable raw material. This is proven by the results of this study: The growth experiments with the marine polychaete *N. diversicolor* showed considerable growth and a superior fatty acid profile in its body tissue.

The synthesis of HUFA and PUFA is well established for freshwater fish (Tocher *et al.* 2004; Hastings *et al.* 2001) but not for marine fish species. However, non-ability cannot be assumed as a norm in marine fish. Linares and Henderson (1991) describe the ability of juvenile turbot *Scophthalmus maximus* to synthesize C 22-PUFA from C 20-PUFA, even if it was not extensive. A significant elongase activity in juvenile turbot was found by Bell *et al.* (1994). Other groups found only limited modifications of dietary fatty acids in turbot; they may be elongated by two carbon atoms (Owen *et al.* 1975). Thus, synthesis is likely not the mayor source of essential fatty acids.

Fatty acids from fish feed were incorporated by the fish. This is obvious from the fatty acid composition of the supplied feed and the fish liver samples. Saturated fatty acids, such as myristic, palmitic and stearic acid were present in the fish feed and in liver samples of cultivated fish in similar fractions. The composition of unsaturated fatty acids in the fish was different from their food. A higher fraction of unsaturated fatty acids were found in the fish livers (20% DHA) compared to the feed (13% DHA). This result is not in agreement with the statement by Ibeas *et al.* 2000 that the fatty acid profile of body tissues corresponds to that of the feed. Fatty acid conversion is well established for freshwater fish but may be crucial in marine fish (Sargent *et al.* 1995). There is no doubt that the majority of fatty acids found in the fish came from the feed. But, the higher fraction of DHA cannot be explained without a synthesis by the fish. C18:3 (n-3), α-linolenic acid and C20:4 (n-6), arachidonic acid were detected in the fish liver but not in the fish feed. Both fatty acids are precursors for the C20:5 (n-3), eicosapentaenoic acid and C22:6 (n-3), docosahexaenoic acid pathway. The fatty acid profiles, i.e. C22:6 > C20:5 > C20:4 > C18:3, of the fish livers give evidence that a conversion involving Δ4, Δ5 and Δ6 desaturases was possible. However, C20:4 (n-6), arachidonic acid may be essential (Linares and Henderson 1991) in *S. aurata*. Tocher and Ghioni (1999) found a deficiency in the Δ5 desaturase in gilthead sea bream *S. aurata*, which is necessary for the synthesis. Alpha-linolenic acid is essential in all vertebrates as they lack Δ12 and Δ15 desaturases which are necessary for the *de novo* synthesis from C18:1(n-9), oleic acid. Thus C18:3 (n-3), α-linolenic acid is by all means essential.

The analyses of materials from the multitrophic integrated recirculating system showed that fatty acids can be recycled. Sediment samples showed marked differences in the fatty acid profile compared to all fish associated samples. Carbon-chain length between 16 and 18 C-atoms were most dominant and only a limited number of unsaturated fatty acids were detected. The most prominent fatty acids were palmitic acid and Alpha-linolenic acid, accounting for more than 50 % of the total fatty acid composition in the sediment. The absence of long chained polyunsaturated fatty acids might be due to β-oxidation by bacteria within the sediment. Another explanation could be the immediate uptake of fatty acids by the worms. From these results it can be concluded that microbial fatty acid synthesis via well described metabolic pathways (Sijtsma and de Swaaf 2004; Russell and Nichols 1999) had not been taken place in sediments. Thus, an upgrade of fish faeces in sediments via an internal microbial loop cannot be elaborated from the results of this study.

N. diversicolor is an omnivorous organism, able to switch its feeding behaviour according to the environmental conditions (Hartmann-Schroeder 1996; Scaps 2002). The possible resources fish faeces, uneaten fish feed and sediment borne bacteria provide different dietary fatty acids to the worms. Therefore it must be assumed that the worms used different feed sources during the experiments, which may be expressed by subsequent changes in body composition.

The samples of *N. diversicolor* showed variations in their fatty acid profiles. Worms caught from the wild contained a lower number of fatty acids compared to cultured individuals. The most dominant fatty acid of the wild caught samples was EPA. Beside EPA, fatty acids with carbon-chain lengths of 16 – 18 C-atoms were most prominent. Saturated and monounsaturated fatty acids accounted for more than half of the total fatty acid composition of wild caught individuals of *N. diversicolor*. The composition of *N. diversicolor* did not change significantly after a short term experiment and fish faeces as exclusive food source. According to Kirsch *et al.* (1998) significant changes in fatty acid profiles needs more than three weeks. Therefore it can be assumed that a ten days experiment was not sufficient to detect an alteration of the fatty acid profile of the worms. In the long-term experiment fatty acid profiles were significantly upgraded. Especially C20:4 (n-6), C20:5 (n-3) and C22:6 (n-3) were available in high fractions. These fatty acids are essential for fish and other aquaculture species.

In natural habitats benthic microbes might be an important source of nutrients for *N. diversicolor* (Harvey and Luoma 1994; Plante and Mayer 1994; Lucas and Bertru 1997). However, as written above, no bacterial upgrade can de derived from the results of this study, even if Lucas *et al.* (2003) describe an alteration of the bacterial communities inside the worm bed (sediment). It remains unclear whether a modified bacterial community in the worm bed may lead to a better nutrient supply to the individual worm. Cannibalism could be a valuable vector for essential nutrients. However, cannibalism, which is quite common in *N. diversicolor* (Hartmann-Schroeder 1996; Batista *et al.* 2003) and may supply essential nutrients, was of minor importance in the experiments of this study. High survival rates were always observed.

The comparison of wild caught and cultured worms revealed that significant differences occurred between different habitats and thus, the available food. This is supported by Fraser *et al.* (1989) and Graeve *et al.* (1994) who used fatty acid signatures as tracers for feeding behaviour during investigations of food webs. Due to the lack of the metabolic pathways, many marine organisms incorporate fatty acids with little or no modifications (Budge *et al.* 2002). From that the authors concluded that fatty acids can be used as indicators of the dietary source. In our case most of the fatty acids found in fish faeces were also found in tissues of *N. diversicolor*. In particular, 16 out of the total 17 fatty acids found in the fish faeces were also found in the worms. Furthermore nine fatty acids found in the faecal material of the fish were also found in the fish feed. Uneaten fish feed may have been an additional source of fatty acids to the worms. This assumption should be further investigated to reveal the nutritional pathway of *N. diversicolor*.

Implications for aquaculture

The importance of high quality raw materials for aquafeeds has already been recognized (Hillestad and Johnson 1994; Wilson and Moreau 1996; Davis 2001). Caused by the intensification of aquaculture production and, in all likelihood, the stagnation of fish meal and fish oil supply alternatives for dietary sources of protein, lipid, amino acids and vitamins are on demand (Olive 1999). Numerous investigations revealed the consequences of lacking one or more essential fatty acids during the culture of marine organisms such as fish and crustaceans. Sargent *et al.* (1999) reported that an insufficiency of DHA in fish larvae is likely to impair neural and visual development. This could have serious consequences for a whole range of physiological and behavioural processes including those dependent on neuroendocrines such as sexual maturation. Worms cultured during this experiment contained significant amounts of DHA. Koven et al. (2001) tested the effects of high dietary DHA, C22:6 (n-3) and varying arachidonic acid (C20:4 (n-6)) on growth, survival and resistance to handling stress in sea bream larvae. Their results suggested that dietary arachidonic acid (AA), fed prior to handling stress, improved the survival more effectively than fed following handling stress, which could be due to the possible effect of AA-based eicosanoids on cortisol production during stress. The dietary AA effect on survival may be related to stress resistance. Polychaetes form a major component of the natural diets of several marine finfish and larger crustaceans (Olive 1999; Shucksmith et al. 2006). Therefore, worms could be an important vector for the transfer of EFAs to cultured fish and crustaceans. Oleic acid, C18:1 (n-9), is described by Dalsgaard et al. (2003b) as the precursor of all (n-3) and (n-6) PUFAs, which are essential to heterotrophic organisms. It seems reasonable to assume that *N. diversicolor* cultured with particulate matter from an aquaculture system contains substantial fractions of this fatty acid although it is not possible with our methods to distinguish between C18:1 (n-9) and C18:1 (n-12). Rodriguez et al. (2004) reported that percentages of AA, EPA and DHA of wild caught black seabream (*Spondyliosoma cantharus*) were considerably higher than in captive fish fed with commercial diets. The absence of spawning in the captive fish suggests that future research on the lipid requirements of this omnivorous species is necessary in order to estimate whether the administration of currently available aquaculture formulated feeds is adequate for a good black seabream performance and reproduction. Bell and Sargent (2003) also stated that eicosanoids derived from AA are physiologically active in fish and that for example series-2-prostaglandines derived from AA have long been used to induce spawning in fish. Furuita et al. (2006) found that the DHA level in the polar lipid fraction of high quality eggs from the Japanese eel (*Anguilla japonica*) were significantly higher than in low quality eggs. It may be assumed that DHA is important for larval development of eel.

Combining the results on the requirements of aquatic organisms for EFAs leads to the conclusion that the fatty acid composition of *N. diversicolor* cultured with particulate matter from aquaculture system could be a serious source of dietary fatty acids in aquaculture. Employing the growth results (see Chapter 3 of this publication) with the fatty acid contents allows estimations about the prospects of *N. diversicolor* as resource for essential raw materials for aquaculture. Artificial wetlands supplied with effluents from aquaculture are able to develop small ecosystems. Applying different conclusions derived from the experimental culture of the worms (e.g. average stocking density, dry matter content and fatty acid content) about 49 kg fatty acids harvestable per hectare wetland can be estimated. Integrating this fatty acid production in a fish farming system with an annual production of 100 tonnes, approx. 13 kg fatty acids can be additionally produced by the culture of detritivorous worms. Nevertheless, Vijayan *et al.* (2005) indicated that polychaete worms act as passive vectors of white spot syndrome virus (WSSV) in the transmission of white spot disease to *Penaeus monodon* broodstocks. Polychaetes form an indispensable component of the maturation diet of penaeid shrimp broodstocks in hatcheries all over the world due to their high nutritive values (Bray and Lawrence 1992). In India, the annual consumption of polychaetes by shrimp hatcheries is estimated to be about 16 to 20 tonnes (Vijayan *et al.* 2005). As there is no polychaete aquaculture in India, the entire polychaete biomass is collected from natural habitats. To overcome the problems of virus infections, developments towards pathogen free broodstocks in biosecure systems have been done in recent years (Lotz 1997; Ogle and Lotz 1998; Fast and Menasveta 2000; Preston *et al.* 2004). These developments included research into specific pathogen free (SPF) and specific pathogen resistant (SPR) shrimp seed, reduced or zero water exchange during pond grow out, development of probiotics and immunostimulants to reduce disease susceptibility and genetic selection and improvements through closed lifecycle culture. Little attention was given to specific pathogen free food sources which could act as vectors for viruses. Worm culture as described here in this research provides the possibility of producing pathogen free diets for shrimp and fish culture.

5. Conclusion

A recycling of valuable feed nutrients such as fatty acids can be achieved in integrated aquaculture and thus, a production of raw materials demanded for animal nutrition in aquaculture is possible. A quantitative and qualitative upgrade of waste nutrients can be achieved by the cultivation of *N. diversicolor*. The absolute amount of fatty acids after the culture within the integrated recirculating system increased by a factor of 1.6 compared to wild caught individuals of *N. diversicolor*. A shift towards long chained, highly unsaturated fatty acids caused by the dietary conditions during the culture with solid waste was evident.

N. diversicolor showed its ability to re-use nutrients, which are otherwise discharged from the production system. The fatty acid profiles of the worms indicate that they fed on solid waste, including faecal material, uneaten fish feed and bacteria.

Nevertheless, fish nutrition within the integrated recirculating system was in excess. Hints towards the ability of marine fish to synthesize essential fatty acids were detected and need to be further investigated to achieve improved diets to meet the physiological requirements of cultured fish.

Future research should also focus on the exact pathways of fatty acid synthesis of benthic ecosystems and its transport pathways through the food chain. Contrary statements about the *de novo* synthesis of fatty acids by bacteria are available and need to be inspected.

Chapter 5

Impact of *Nereis diversicolor* (O.F. Mueller, 1776) on nitrification and nitrifying bacteria in two types of sediments

Bischoff A.A. and Prast M.

Abstract

Nitrification is a microbial process which is catalyzed by bacteria. While numerous autecological studies on these nitrifying bacteria are published, only few publications consider bacteria involved in the nitrification process as an important part of the benthic microbial food web. We tested the hypothesis that the polychaete *Nereis diversicolor*, described as an organism able to feed on bacteria, might influence the nitrification potential. The effects can be direct by actively changing the bacterial community or indirect by bioturbation and therefore enlarging the surface layer and supplying substrate to the bacteria. Experiments with two types of sediments in laboratory flumes, with and without the addition of *N. diversicolor*, were conducted. Higher bacterial abundances in the presence of worms and changes in nitrifying bacteria between treatments for the fine sediment were found. The grain size distribution of the sediment seems to have an impact on the nitrification potential and abundance of bacteria.

1. Introduction

The nitrogen cycle is one of the most important biogeochemical cycles as nitrogen is an essential nutrient for all organisms. Various types of prokaryotes are responsible for transformations within the nitrogen cycle such as nitrogen fixation, nitrification, nitrate reduction or denitrification (Ramaiah 2005). A lot of studies dealing with these bacteria have been conducted in the past, but only few considered them as part of a food web, in which these bacteria have to compete for nutrients/substrates with other organisms, such as benthic microalgae or ammonia oxidizing crenarchaeota (Risgaard-Petersen *et al.* 2004; Koenneke *et al.* 2005), or in which they might be a prey for other animals (Verhagen and Laanbroek 1992).

This study focussed on nitrifying bacteria in sediments of the Baltic Sea and their interaction with a benthic macroinvertebrate, the polychaete *N. diversicolor*. Nitrification is the sequential oxidation of ammonium (NH_4^+) to nitrite (NO_2^-) and nitrate (NO_3^-) and is predominantly accomplished by chemolithotrophic bacteria under oxic conditions. This process is a dominant metabolic pathway in the water column and the upper layer of aquatic sediments. The nitrifying bacteria consist of two functionally different groups of proteobacteria: ammonium oxidizing bacteria (AOB) and nitrite oxidizing bacteria (NOB) (Spiek and Bock 1998). The ammonia oxidizing crenarchaeota (Koenneke *et al.* 2005) were only discovered during the course of this study and, thus, were not taken into account, but they should be considered for further studies.

Deposit feeding organisms, such as polychaetes are described as important bacterial grazers in aquatic sediments (Grossmann and Reichardt 1991; Sherr and Sherr 2002; Lucas et al. 2003). Plante et al. (1989) concluded that grazing pressure of polychaetes had an effect on bacterial activity and alternates the community structure in the sediment. Hence polychaete grazing might also affect the metabolic pathways in which bacteria play an essential role. Polychaete species are known as important bioturbator of shallow sediments and it is an indigenous species of the sediments used during the experiment. *N. diversicolor* inhabits sediments such as sandy mud or gravel and is able to adapt to a wide variety of environmental conditions, including salinity, temperature and oxygen availability.

The aim of the present study was to find out whether a macroinvertebrate in-fauna representative could have an impact on the nitrification in sediments by grazing pressure and/or by bioturbation. A laboratory experiment was conducted using flumes with natural sediments with and without the addition of *N. diversicolor* to analyse the effects on the abundance and taxonomic composition of nitrifying bacteria in the sediments and their nitrification potential.

2. Material and Methods

2.1 Experimental Setup

Experiments were conducted in four laboratory flumes (length: 32.4 cm, width: 17.4 cm, height: 18.7 cm, depth of sediment layer ~7.0 cm). Natural sediments were collected from the Baltic Sea (Bay of Kiel, approximately 54°21′N 10°09′E) and after cleaning and drying for 48 hours at room temperature incubated in the flumes for 30 days prior to the stocking of the worms. Two types of sediment were used. Sediment 1, the "fine sediment", had a grain size of 0 – 2 mm and sediment 2, the "coarse sediment", had a grain size of 0 – 4 mm (Tab. 1). Water replacement per flume was ~10 L h^{-1} with artificial sea water (salinity ~29 psu, temperature ~20 °C). All flumes were connected to form a closed recirculating system which also included a basin for water collection, a biofiltration unit and a pump for water circulation. Light regime was 12:12 h light/dark with artificial light (~25 µE m^{-2}s^{-1}).

Tab. 1: Grain size distribution of both sediment types. Figures display means ± standard deviation (n = 6). For the fine sediment the major fraction was in the range between 0.49 and 0.125 mm, whereas for the coarse sediment the largest part was above the range of 0.5 mm.

Grain size [mm]	% portion	
	fine sediment	coarse sediment
> 0.5	32.2 (± 3.7)	96.6 (± 0.9)
0.49 – 0.125	63.1 (± 3.4)	0.4 (± 0.3)
0.125 – 0.063	1.1 (± 0.6)	0.0 (± 0.0)
0.063 – 0.025	0.3 (± 0.3)	0.1 (± 0.0)
< 0.025	3.3 (± 0.7)	2.9 (± 0.6)

All worms used during this experiment were first generation offspring, reared from wild caught individuals, in a laboratory recirculating system. They were chosen randomly and transferred to one flume of each sediment type. The flumes stocked with worms were referred to as +worm treatments. The second flume of each sediment type was used as a control in the absence of *N. diversicolor* and these flumes were referred to as –worm treatments. Both treatments with worms were stocked 40 days prior to the measurements with 30 individuals per flume resulting in calculated initial densities of approximately 1000 Individuals m^{-2}, which is according to Davey (1994) close to natural densities of adult *N. diversicolor*.

2.2 Sampling procedure

All samples of the fine sediment were taken with a plastic sediment corer (∅ 1.6 cm, total length 8.2 cm) by drilling the corer into the sediment, sealing the upper end and pulling the filled corer out of the sediment. Samples for total prokaryotic abundance were transferred into pre-weighed 50 ml cetrifugation tubes and were fixed with glutardialdehyde (GA; 2% f.c.). Samples for the analysis of nitrifying bacteria were transferred into a second set of 50 ml centrifugation tubes and were fixed with paraformaldehyde (4% f.c.). Of each of the four flumes used for the experiment three samples from different locations per flume were collected. During the complete acclimatisation phase no worm mortality was observed. Sampling locations were selected randomly not considering the location of worm burrow entrances assuming a complex net of burrows within the entire sediments as described by Davey (1994).

Samples from the "fine sediment" for the determination of the nitrification potential were collected with the same type of corer but transferred directly into Erlenmeyer flasks used for a slurry assay (see below) with six replicates per flume. Samples from the "coarse sediment" were collected with a spoon and processed analogously to the samples from the fine sediment. The change of sampling procedure was necessary because it was not possible to collect sufficient amounts of coarse sediment with the sediment corer used for the fine sediment. Samples collected with the spoon included all depth horizons of the sediment.

All samples for the determination of abiotic parameters such as porosity, ash free dry weight and grain size were also collected with a spoon. Wet weights of sample material were recorded before the sediment samples were dried at 60 ± 5 °C for 24 hours. Six replicates per flume were sampled.

2.3 Prokaryote counts

Numbers for total bacterial abundances were obtained using the DAPI-method (Porter and Feig 1980). Samples were sonicated in order to remove attached bacteria from sediment grains (Labsonic M, puls: 0.9, amplitude: 60%, 60 sec.). Samples were resuspended in sterile artificial seawater with 2% GA and gently shaken by hand for 60 seconds. Afterwards, 3.5 ml of supernatant was immediately pipetted off into a 50 ml centrifuge tube. To achieve a quantitative bacteria removal from particles, this washing procedure was repeated at least five times per sample. 200 or 250 µl of the supernatant, diluted in 2 ml of particle free artificial seawater, were then filtered onto a black 0.2 µm-filter (Nuclepore) for prokaryotic counts.

The NOB *Nitrospira* sp. and *Nitrobacter* sp. as well as the ammonium oxidizing β-Proteobacteria (β-AOB) were detected with fluorescence-*in-situ*-hybridization (FISH). Probes Ntspa712 and cNtspa712 (Daims *et al.* 2001), NIT3 (Wagner *et al.* 1996) were used for NOB and Nso1225 (Mobarry *et al.* 1996) for β-AOB, respectively (for details on these oligonucleotide probes see probeBase; (Loy *et al.* 2003)). All probes were labeled with Cy3. Hybridization procedure followed the protocol by Pernthaler *et al.* (2001). Washing and filtration was identical to the procedure applied for the DAPI-method except that white 0.2 µm-filters (Nuclepore) were used. All counts were conducted with a Nikon Eclipse E800 microscope.

2.4 Nitrification potential (Slurry assay)

Ammonia oxidizing potential and nitrite oxidizing potential were analysed by performing two different approaches. A total number of twelve Erlenmeyer flasks for each oxidizing approach were used. This total number was divided into different groups such as sediment types, "fine" and "coarse", and sediments stocked with worms and sediments without worms. For each set three replicates were analysed for each sampling time.

Sediment samples were slurried by adding filtered artificial seawater (~6.5 g sediment + 100 ml 0.2 µm-filtered artificial seawater per flask) and placed into 200 ml Erlenmeyer flasks. All flasks of the ammonia oxidizing group were supplied with NH_4^+ (50 µM) and $NaClO_3$ (20 mM) for analysing the NH_4^+-oxidation rates. All flasks of the nitrite oxidising group were supplied with NO_2^- (25 µM) and allylthiourea (10.0 mg/l) for analysing the NO_2^--oxidation rates. The Erlenmeyer flasks were aerated via an air pump and stirred with a magnetic stirrer. Samples were taken out of each flask after 0, 1.5, 5 and 22 h (t_0, t_1, t_2, t_3). Prior to collecting one sample of 10 ml per flask by using a pipette, stirring and aeration were stopped for five minutes. Samples were centrifuged for five minutes (Heraeus Labofuge 200, 5000 rpm) to remove remaining sediment particles. NO_2^- concentrations were measured photometrically using a Bran + Luebbe AutoAnalyzer (AA3) (Dollhopf et al. (2005), Stief pers. comm.).

2.5 Abiotic parameters

Temperature, pH, oxygen saturation and salinity in the water circuit were measured with a multiprobe (WTW Multimeter 350i; Tab. 2), NO_2^- in situ concentrations were measured photometrically (Bran+Luebbe, AutoAnalyzer AA3). Dry weight, total organic matter (TOM), porosity (calculated according to Hölting 1996) and grain sizes were determined (Retsch AS 200 basic wet sieving machine). All results given in g^{-1} referred to dry weight. SigmaXL (Version 2000) and SPSS (Version 11.0) software were used for statistical analysis. Data were tested for normality (Kolmogorov-Smirnov test) and for homogeneity of variances (Levene's test) prior to analysis. t-tests and analyses of variances (ANOVA) which were followed by Tukeys HSD posthoc tests were performed with the data.

3. Results

Except sediment porosity all other abiotic parameters did not differ significantly between treatments (Tab. 2).

Tab. 2: Abiotic parameters (means ± SD) measured in situ within the experimental flumes. For pH, oxygen saturation, temperature and salinity n = 5 and the number of replicates for sediment porosity and TOM was n = 3.

	fine sediment		coarse sediment	
	+ worms	- worms	+ worms	- worms
pH	8.15 ± 0.09	8.18 ± 0.06	8.18 ± 0.05	8.19 ± 0.05
oxygen saturation [%]	83.0 ± 5.8	82.8 ± 5.8	88.3 ± 3.4	88.5 ± 3.6
Temperature [°C]	19.4 ± 0.3	19.4 ± 0.3	19.4 ± 0.3	19.4 ± 0.3
Salinity [psu]	29.5 ± 0.6	29.5 ± 0.6	29.5 ± 0.6	29.5 ± 0.6
Sediment porosity	24.4 ± 0.7	25.5 ± 0.6	18.4 ± 1.4	19.1 ± 2.7
TOM [%]	0.22 ± 0.06	0.20 ± 0.03	0.32 ± 0.04	0.37 ± 0.05

3.1 Bacteria

Over all treatments, total bacterial abundances ranged from 9.4×10^7 to 2.5×10^8 cells g^{-1}. Numbers differed significantly between treatments for the fine sediment with and without worms (Tab. 3; Fig.1). The bacterial abundance for the treatment with worms present was on average almost twice as high as the treatment without worms. For the coarse sediment the difference between treatments was statistically not significant. For this sediment in the +worm treatment one bacterial count showed lower values compared to the other two counts.

Comparing the two types of sediment bacterial abundance differed significantly in the absence of polychaetes (Tab. 3; Fig.1) showing higher values for the coarse sediment.

Tab. 3: t-test results from total bacterial abundances in the two types of sediment with and without worms.

	Levene's test		t-test		
	F	Sig.	df	t	Sig. (2-tailed)
fine sediment					
+worms vs. –worms	5.778	0.074	4	6.651	**0.030**
coarse sediment					
+worms vs. –worms	8.388	0.044	4	0.649	0.552
worms present					
fine sediment vs. coarse sediment	5.489	0.079	4	0.362	0.735
worms absent					
fine sediment vs. coarse sediment	0.136	0.731	4	-7.616	**0.002**

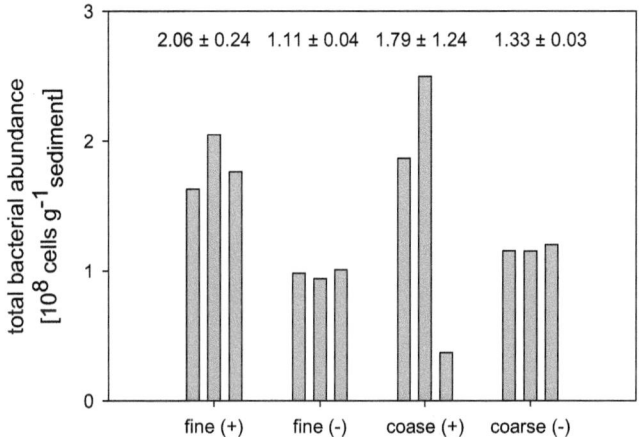

Fig. 1: Total bacterial abundances obtained from all treatments. Fine and coarse represent the type of sediment used during the experiment and (+) and (-) represent the presence or absence of *N. diversicolor*. Bars present the abundances counted for each replicate. Figures displayed above the bars represent the means ± standard deviations of the corresponding treatment.

Abundances of the three investigated types of nitrifying bacteria (*Nitrospira* sp., *Nitrobacter* sp. and β-AOB) differed among treatments: In the fine sediment abundances were lower in the treatment without polychaetes but in the coarse sediment abundances were higher in the treatment without polychaetes. Combined the investigated nitrifying bacteria contributed 4.65 and 3.80% of total bacteria in both treatments of the fine sediment. For the coarse sediment they accounted for 1.85 and 2.08% in both treatments, respectively. The abundances of the three types of nitrifying bacteria differed within the treatments (Fig. 2) and for *Nitrospira* in the fine sediment this difference was statistically significant (Tab. 4; Fig. 2). The total abundance of *Nitrospira* sp. increased in the presence of *N. diversicolor*, as well as the relative abundance did. The total and relative abundance of *Nitrobacter* sp. doubled in the presence of the worms. The situation for β-AOB was different. In the absence of the worms total and relative abundance of these bacteria were significantly higher (Tab. 4; Fig. 2). For the coarse sediment no differences were detected.

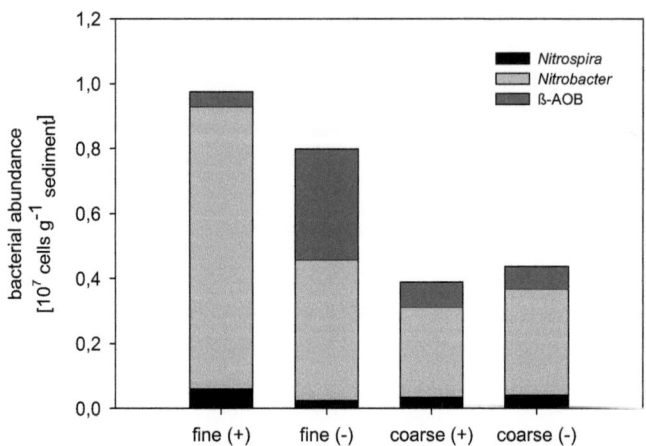

Fig. 2: Nitrifying bacteria detected by FISH. For *Nitrospira* Ntspa 712 and cNtspa 712 probes were used. The probes NIT3 and Nso1225 were used for *Nitrobacter* and β-AOB, respectively. Fine and coarse represent the type of sediment used during the experiment and (+) and (-) represent the presence or absence of *N. diversicolor*.

Tab. 4: One-way ANOVA results from the nitrifying bacteria.

	df	MS	F	Sig.
Nitrospira				
Between groups	3	0.067	4.854	**0.033**
Within groups	8	0.014		
Total	11			
Nitrobacter				
Between groups	3	21.794	2.163	0.170
Within groups	8	10.076		
Total	11			
ß-AOB				
Between groups	3	5.784	8.219	**0.008**
Within groups	8	0.704		
Total	11			

3.2 Nitrification potential

For both sediments nitrification potentials were measured at different time intervals and the rates $t_0 - t_3$ as results of linear regression are presented.

3.2.1 Fine sediment

Nitrification potential rates for the fine sediment ranged from 0 to 0.63 µg N g^{-1} h^{-1} (Fig. 3).

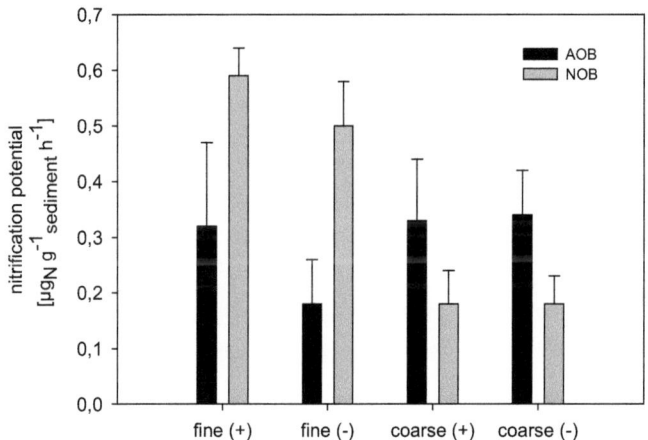

Fig. 3: Nitrification potentials (AOB = ammonia oxidizing bacteria; NOB = nitrite oxidizing bacteria) Fine and coarse represents the type of sediment used during the experiment and (+) and (-) represent the presence or absence of *N. diversicolor*. Bars present means + standard deviation (n = 3).

In the treatment without worms the rates were on average lower than in the treatment with worms. Rates for ammonia oxidation were 0.32 ± 0.15 µg N g^{-1} h^{-1} for +worm treatments and 0.18 ± 0.08 µg N g^{-1} h^{-1} for –worm treatments, respectively. Nitrite oxidation rates were 0.59 ± 0.05 µg N g^{-1} h^{-1} for +worm and 0.50 ± 0.08 µg N g^{-1} h^{-1} for –worm treatments, respectively. No statistical differences for ammonia and nitrite oxidation between the treatments with and without worms were detected.

3.2.2 Coarse sediment

For the analysed coarse sediment the nitrification potentials ranged from 0.14 to 0.45 µg N $g^{-1}h^{-1}$ (Fig. 3). Ammonia oxidation rates were higher compared to nitrite oxidation rates. Rates for ammonia oxidation ranged from 0.22 to 0.45 µg N g^{-1} h^{-1} and were 0.33 ± 0.11 µg N g^{-1} h^{-1} for +worm treatments and 0.34 ± 0.08 µg N g^{-1} h^{-1} for –worm treatments, respectively. Nitrite oxidation rates ranged from 0.14 to 0.25 µg N g^{-1} h^{-1}. Rates were 0.18 ± 0.06 µg N g^{-1} h^{-1} for +worm and

0.18 ± 0.05 µg N g^{-1} h^{-1} for –worm treatments, respectively. No significant differences for ammonia and nitrite oxidation between the treatments with and without worms were detected.

4. Discussion
4.1 Bacterial abundance

Total bacterial abundances were well within the range reported for other marine and freshwater sediments (Llobet-Brossa *et al.* 1998; Kuwae & Hosokawa 1999, Altmann *et al.* 2004). The presence of *N. diversicolor* did obviously enhance the bacterial abundance in both types of sediments, though the differences on flumes without worms were statistically significant for the fine sediment only. This was due to high variation of bacterial numbers in the coarse sediment. Including all data into the analyses for total bacterial abundances of the coarse sediment led to no statistically significant differences between treatments with and without worms. Considering the low count of total bacterial abundances as an outlier would change the situation severely.

It could be expected that due to dense sediment layers, caused by fine sediments, dissolved nutrients are only available at the sediment surface. Caused by an enhancement of surface area (Fenchel 1996) and the active water movement caused by the worms (Riisgard *et al.* 1996) nutrient as well as oxygen availability for bacteria should be increased especially in fine sediments. Kristensen (1984) suggested nutrient recycling by bacteria for ammonia excreted by *Nereis virens*. Such a nutrient recycling was also reported by Reichardt (1988) for psychrophilic bacteria. Predation pressure of *N. diversicolor* on bacteria might further keep bacteria in the exponential growth phase and consequently enhance bacterial abundance. Nevertheless, no statistically significant effect of the sediments was revealed.

4.2 Taxonomic composition of nitrifying bacteria

In some studies bacteria of the genus *Nitrospira* have been identified as the dominating nitrite oxidizing bacteria (NOB) in marine and freshwater systems (Hovanec *et al.* 1998, Altmann *et al.* 2003). In contrast, in our study bacteria of the genus *Nitrobacter* were the most important NOB, differing significantly from *Nitrospira* by factors of more than 14 and 16 in the fine sediment and more than 7 in the coarse sediment. In total they accounted for 50 – 89% of all nitrifying bacteria. Other NOB such as *Nitrospina* or *Nitrococcus* might have also been present, but are unlikely to be numerically important (Watson et al. 1981). The β-AOB were found in significantly higher abundances in the absence of worms for the fine sediment. This is unexpected considering the ammonia excretion of the worms implying increased nutrient availability in the presence of the worms. However, the opposite was found in the fine sediment. The β-AOB are gram-negative as most of the sediment surface bacteria (Moriarty and Hayward 1982). Lucas and Bertru (1997) described lytic activity for gram-negative bacteria in extracts of the digestive tract of *N. diversicolor*. Plante and Mayer (1994) stated that deposit-feeders may grow on a diet of bacteria. Grossman and Reichardt (1991) showed that the digestive removal of bacterial carbon could attain 95% for *Arenicola marina*. Harvey and Luoma (1994) supposed that organic matter gained from bacteria such as exopolymeres and cellular envelopes may be of trophic importance. Therefore, it could be assumed that the feeding and digestion of β-AOB by *N. diversicolor* could be the explanation for the reduced abundance of ammonia oxidizing bacteria and therefore the alteration of the nitrifying bacterial community. In contrast to this assumption is that *Nitrospira* sp. and *Nitrobacter* sp. are also gram-negative bacteria and therefore the proportions of bacteria should have been similar. The location of the β-AOB could be a possible explanation, so that the bacteria could be more accessible for the worms.

4.3 Nitrification potential

The increased numbers of nitrifying bacteria in the presence of *N. diversicolor* for the fine sediment should also have led to increased nitrification potentials. Significantly higher ammonia and nitrite oxidation rates could not be detected. This is in contrast to findings of Kristensen (2000) and Heilskov and Holmer (2001). Yingst and Rhoads (1981) suggested that feeding of bioturbating invertebrates on the microbial community in their close vicinity keep these bacteria in an active physiological state, explaining the elevated nitrification potentials in the presence of *N. diversicolor* for the fine sediment. For the coarse sediment the ammonia oxidation potential exceeded the nitrite oxidation potential. These findings are in contrast to natural sediments where higher nitrite oxidation potentials compared to ammonia oxidation potentials are commonly found (Schwoerbel 1999). Interactions of denitrification with other processes of the nitrogen cycle such as anammox could be an explanation. According to Dalsgaard et al. (2003a) the requirements for anammox, the anaerobic bacterial oxidation of ammonia with nitrite, are nitrate rich waters and anoxic conditions. These conditions might occur within the flumes and may have reduced nutrients available for nitrite oxidising bacteria. The reduced nutrient availability could have led to nitrite oxidising bacteria at a low active physiological state. Alteration of this physiological state can only be achieved by increased nutrient concentrations over a certain time period and the time available for the slurry assays was too short to complete this change. Syakti et al. (2006) reported lag phases of 2 days for two strains of bacteria and Dean (1957) showed that bacteria which remained longer in liquid media had a decreased lag phase on plates.

5. Conclusions

N. diversicolor causes a bioturbation effect by its burrowing behaviour which increases the total bacterial abundance. It may alter the nitrifying bacterial community and hence the nitrification potential of sediments inhabited by the rag worms is increased. This effect is influenced by the type of sediment. Bioturbation caused by worms increases nitrification potential until a certain grain size is exceeded, afterwards this effect diminish. The presented study concentrated on the effects caused by macrobenthic organisms on bacterial abundances, community structures and nitrification potentials, further research is required to reveal the complex influence of N. diversicolor on benthic microbial food webs such as ingestion and digestion of different bacterial strains.

Chapter 6

Cultivation of the European brown shrimp (*Crangon crangon*) to evaluate its potential as secondary organisms for integrated aquaculture systems fed with solid waste

Bischoff A.A., Hielscher N., Marohn L. und Waller U.

Abstract

Solid waste from aquaculture productions is in the range of 10 to 15% of the applied amount of food. Its biochemical composition is still sufficient for the culture of organisms which are located low in the food chain, such as detritivorous organisms.

This chapter combines the results of two experiments performed to evaluate the potential of the brown shrimp (*Crangon crangon*) as an organism for integrated multitrophic aquaculture. Two small scale recirculating systems were used for the cultivation of *C. crangon* which were fed with solids originating from fish culture. Criteria for the evaluation of the crustacean as a candidate for integrated aquaculture were growth and survival as well as the biochemical composition of the animals.

Hints towards farming *C. crangon* as secondary organisms in multitrophic integrated aquaculture fed with solids were detected. Specific growth rates up to 0.006 d^{-1} could be achieved. Survival rates of *C. crangon* during both experiments ranged between 20 and 75%. Biochemical composition of cultivated shrimps improved with increasing amounts of food and increasing quality of the applied food.

1. Introduction

Global finfish aquaculture increased more than 4% in 2004 and the increase of total aquaculture production in 2004 was more than 7% (FAO 2006a). This led to increasing amounts of dissolved and particulate waste released by organisms produced in aquaculture. Solids originating from open aquaculture systems generally consist of faecal material and uneaten food particles. In recirculating aquaculture systems an additional source of solids are biofilms growing on surfaces such as tanks or biofilters (Franco-Nava *et al.* 2004). Between 14 and 23% of the excreted nutrient waste is released as particulate waste.

Solids originating from commercial food applied in aquaculture have in common that their biochemical composition (organic and energy content, carbon and nitrogen concentrations) is still sufficient for the culture of organisms which are lower in the food chain. In natural systems nutrients are cycling through food webs. The technical application of simple food webs in aquaculture is called integrated multitrophic aquaculture (Chopin *et al.* 2006) or integrated aquaculture (Neori *et al.* 2004; Schneider *et al.* 2005; Wecker 2006). Research so far focussed mainly on the utilisation of dissolved inorganic nutrients as they amount for the larger fraction of released waste. Only few authors deal with solids derived from aquaculture as resources for further use by organisms located at lower trophic levels (Ryther 1983; Batista et al. 2003; Milanese et al. 2003; Erler et al. 2004).

This chapter combines the results of two experiments performed to evaluate the potential of the brown shrimp (*C. crangon*) as a potential candidate for integrated multitrophic aquaculture in recirculating systems. Two small scale recirculating systems were used for the cultivation of *C. crangon*, which were fed with solids collected from a recirculating aquaculture system used for the culture of Gilthead sea bream (*Sparus aurata*). The concentrations of dissolved inorganic nutrients (Total Ammonia Nitrogen = TAN, nitrite, nitrate and phosphate), total organic contents of the sediment and the abiotic parameters of the water were used to describe the quality of the system during the experiments. Criteria for evaluating the performance of the crustacean as a candidate for integrated aquaculture were growth and survival as well as the biochemical composition of the animals.

2. Material and Methods
2.1 Measurements of abiotic parameters
During the experiments abiotic factors such as temperature, salinity, pH values and oxygen saturation were measured with handheld measuring devices (WTW Multi 350i in combination with CellOx 325 and SenTix 41-3 probes) on a daily basis.

2.2 Analytical procedures
During both experiments with brown shrimp (*C. crangon*) different analytical procedures and methods were applied. These methods included the analysis of dissolved inorganic nutrient concentrations, biochemical analysis of the food sources and animals, as well as growth and survival studies of the animals.

2.2.1 Dissolved inorganic nutrients
Water samples for dissolved inorganic nutrients were collected on a daily basis from all tanks and stored immediately at -20 °C until analysis. Concentrations of dissolved inorganic nutrients such as Total Ammonia Nitrogen (TAN), nitrite-N (NO_2^--N), nitrate-N (NO_3^--N) and orthophosphate (PO_4^{3-}-P) were analysed using a continuous flow analyzer (Bran+Luebbe AA3, Norderstedt, Germany). TAN was analysed by using the salicylate method, nitrite-N by using the diazotation method, nitrate-N by using

the hydrazine reduction method and orthophosphate-P was analysed by using the amino acid method (Grasshoff et al. 1999).

2.2.2 Water content of materials

Water content of the different samples from the commercial fish food, the faecal material, which was applied as food source, and animal tissues from investigated animals were calculated on different intervals.

The water content of all materials was calculated as the weight difference of fresh and dried material according to Winberg and Duncan (1971). For that purpose all collected samples were dehydrated in an oven at 60 ± 5 °C for 24 hours. After dehydration all fractions of total organic and energy contents as well as carbon and nitrogen concentrations refer to dry weights of the sampling material.

2.2.3 Total organic matter content of materials

Total organic matter (TOM) content of collected sample material was determined according to Winberg and Duncan (1971) by calculating the difference of dried sample material before and after applying a muffle furnace at 450 ± 50 °C for 24 hours.

2.2.4 Energy content of materials

For analysing the energy content of the samples, material was dehydrated at 60 ± 5 °C for 24 hours. The dried material was grinded, weighted and used for total combustion within an IKA C4000 Calorimeter (IKA, Staufen, Germany).

2.2.5 Carbon and nitrogen content of materials

Carbon and nitrogen concentrations were determined using gas chromatography (GC) in an element analyzer (EURO EA elemental analyser, Milano, Italy). All samples were dehydrated, grinded and weighed into tin capsules.

2.2.6 Growth of *Crangon crangon*

C. crangon were caught at the North Sea coast near Büsum, Schleswig-Holstein (approximately 54°07'38N and 08°51'32E) and transferred to the laboratory. The shrimps were acclimatised in their transport bags for several hours and afterwards transferred into holding tanks, which were in the vicinity of the recirculating system, and kept their for five days. Each day approximately 80 % of the water volume from the holding tanks was replaced with fresh seawater to keep the water quality as good as possible. Each holding tank was also equipped with an individual air supply. At the start of the experiments 15 animals for each individual culture unit were selected randomly from the complete catch, weighed (initial wet weight) and transferred to the tanks. For the weighing procedure shrimps were put on paper tissues to remove adherent water and located into petri dishes to prevent escape of the animals.

After an experimental duration of 21 days, all surviving animals were recaptured and weighed again (final wet weight). Weight gain was calculated as difference between initial and final average weight. Specific growth rates µ $[d^{-1}]$ were calculated according to Jørgensen (1990) following the equation:

$$\mu = ln(W_I/W_F)t^{-1} \qquad (1)$$

Where W_I and W_F are initial (W_I) and final (W_F) mean body mass (wet weight) and t number of experimental days.

2.2.7 Statistical analyses

All statistical analyses were performed using SPSS software. Data were tested for normality (Kolmogorov-Smirnov test) and for homogeneity of variances (Levene's test) prior to analysis. Analyses of variances (ANOVA) were followed by a Tukey post hoc test for multiple comparisons.

2.3 Experimental set up and design

The first experiment was designed with three treatments and three individual growth units per treatment. Therefore three tanks (water volume 210 Litres tank^{-1}) were subdivided into three sections during the first growth experiment of *C. crangon*. Each tank was used for one of the three treatments. The applied treatments were different amounts of food during the experiment. No temperature regulation was included into the recirculating system.

The second growth experiment was designed with four treatments and four replicates per treatment. It was carried out in a small scale laboratory recirculating system and the system consisted of 16 small tanks (each 36.0 x 19.3 x 20.8 cm; total tank volume 12 L) for the shrimp culture, a sump for water collection, a submerged nitrifying biofilter and a pump. All tanks were connected to individual air supplies to secure sufficient oxygen saturation within each tank. The recirculating system was located in a temperature controlled laboratory to sustain constant temperatures around 10 °C. For more details on recirculating systems see Waller (2000). Tab. 1 presents all treatments applied during the experiments.

Tab. 1: Description of all treatments applied during the experiments with *C. crangon*.

	Treatment 1	Treatment 2	Treatment 3	Treatment 4
Experiment 1				
Food source	None	Solids	Solids	---
Amount of food	0%*	7.5%*	15.0%*	---
Stocking	15 Ind. tank^{-1}	15 Ind. tank^{-1}	15 Ind. tank^{-1}	---
Growth units	3	3	3	---
Experiment 2				
Food source	Solids	Solid	Fish feed	None
Amount of food	20%*	40%*	10%*	0%*
Stocking	15 Ind. tank^{-1}	15 Ind. tank^{-1}	15 Ind. tank^{-1}	15 Ind. tank^{-1}
Replicates	4	4	4	4

*This fraction relates to the initially stocked biomass

During the first experiment treatment 1 was without the addition of food; treatment 2 was supplied with a medium amount of food (equivalent to approximately 7.5% of stocked shrimp biomass d^{-1}). Treatment 3 was supplied with a higher amount of food (equivalent to approximately 15% of stocked shrimp biomass d^{-1}). For the two treatments with food supply, the solids were originating from a recirculating system for the culture of Gilthead sea bream (*S. aurata*). These solids consisted of fish faeces, uneaten fish feed and biofilms from the system.

For the second experiment four tanks were used for each treatment. The treatments differed in the amounts of given food. Fish feed and solids originating form a recirculating system for the culture of Gilthead sea bream (*S. aurata*) were used as food sources. Treatment 1 was supplied with a medium amount of solids (equivalent to approximately 20% of stocked shrimp biomass d^{-1}), treatment 2 was supplied with a high amount of solids (equivalent to approximately 40% of stocked shrimp biomass d^{-1}), treatment 3 was given a low amount of fish feed (equivalent to approximately 10% of stocked shrimp biomass d^{-1}) and treatment 4 was without the addition of food.

2.4 Experimental duration

Both recirculating systems were allowed to develop an active microbiology for a time span of four weeks. Experimental duration was 21 days, starting with the stocking of the tanks and finishing after applying food for 21 days. The first experiment was conducted between the 1st and the 20th of June 2005 and the second experiment between the 2nd and the 22nd of April 2006.

3. Results

3.1 Abiotic parameters

Abiotic parameters recorded during the experiments (temperature, salinity, pH values and oxygen saturation) are presented in Tabs. 2 and 3.

For the first experiment values for temperature, salinity, pH and oxygen saturation showed little variation. Temperature was analogous to room temperature and ranged around 20.5 °C. Salinity was around 22.4 and pH values between 8.7 and 9.1 were recorded, with mean values around 8.8. Oxygen saturation always exceeded 90%.

During the second experiment values for temperature and pH were lower compared to the first experiment and were stable around 10.8 °C. pH values between 8.1 and 8.4 were recorded. Salinity varied between 27.8 and 31.3 with mean values around 29.5 and oxygen saturation was always above 40%, which was lower compared to the first experiment.

Tab. 2: Abiotic parameters (means ± SD) for the first experiment with *C. crangon*. (n = 60)

	Treatment		
	1	2	3
temperature [°C]	20.49 ± 0.56	20.51 ± 0.57	20.51 ± 0.57
salinity [psu]	22.42 ± 0.16	22.43 ± 0.16	22.42 ± 0.14
pH	8.81 ± 0.10	8.81 ± 0.11	8.82 ± 0.11
oxygen saturation [%]	> 90%	> 90%	> 90%

Tab. 3: Abiotic parameters (means ± SD) for the second experiment with *C. crangon*. (n = 80)

	Treatment			
	1	2	3	4
temperature [°C]	10.90 ± 0.20	10.68 ± 0.22	10.88 ± 0.24	10.99 ± 0.24
salinity [psu]	29.47 ± 1.43	29.65 ± 1.43	29.54 ± 1.38	29.43 ± 1.44
pH	8.22 ± 0.05	8.22 ± 0.03	8.23 ± 0.04	8.23 ± 0.07
oxygen saturation [%]	> 40%	> 40%	> 40%	> 40%

3.2 Dissolved inorganic nutrients

The results of the analysed dissolved nutrients of both experiments are displayed in Tab. 4. Values were constantly low, but concentrations during the second experiment were on average higher compared to the first experiment. TAN and nitrite-N showed low values in both experiments, with concentrations not exceeding 0.35 mg L^{-1} (TAN) and 0.05 mg L^{-1} (nitrite-N), respectively. Nitrate-N and orthophosphate-P were also low, with concentrations constantly below 1.5 mg L^{-1} (nitrate-N) and 5 mg L^{-1} (orthophosphate-P).

Tab. 4: Dissolved inorganic nutrient concentrations obtained in the system waters during both experiments with *C. crangon*. All concentrations are displayed in [mg L^{-1}].

	Experiment 1	Experiment 2
TAN	< 0.15	< 0.35
NO_3	< 0.50	< 1.50
NO_2	< 0.05	< 0.03
PO_4	< 5.00	< 5.00

3.3 Biochemical composition of applied food sources

The biochemical composition includes the water, total organic matter and energy content, as well as carbon and nitrogen concentrations of the food sources (Tab. 5).

Tab. 5: Biochemical composition of all food sources used during both experiments with *C. crangon*. Values displayed represents means ± standard deviations. Solids Exp. 1 and Solids Exp. 2 represent the solids used during experiment 1 and experiment 2, respectively. (n ≥ 3)

	Solids Exp. 1	Solids Exp. 2	Fish feed
Water content [%]	94.63 ± 1.31	93.15 ± 0.20	~ 6.40
Organic content [%]	64.57 ± 3.99	49.42 ± 0.58	78.10 ± 2.93
Energy [kJ g_{DW}^{-1}]	14.64 ± 0.50	9.99 ± 0.24	23.36 ± 0.08
Carbon content [%]	36.72 ± 1.53	26.24 ± 0.33	50.20 ± 1.03
Nitrogen content [%]	3.45 ± 0.36	1.95 ± 0.09	7.90 ± 0.19

Water content of the different food sources differed severely – fish feed had a low water content whereas water content of the solids was high, reaching values above 90%. Total organic matter content of solids used during the second experiment was the lowest, followed by the solids used for the first experiment and fish feed, used during the second experiment, with values ranging from approximately 50 to 78% for all samples. Energy of fish feed was more than twice compared to solids used during experiment 2. Energy content of solids used for both experiments varied between experiments. Carbon concentration of fish feed showed the highest values, followed by solids of experiment 1 and the lowest values were observed from the solids used during experiment 1. Nitrogen concentrations followed the same order as carbon concentrations.

3.4 Total organic matter content of the sediment

During both experiments the total organic matter (TOM) content of the sediment was measured at least on a weekly basis. During the first experiment, concentrations showed values around 0.5% and difference between treatment 1 and 3 was significant (Fig. 1a, ANOVA, $p < 0.001$). During the second experiment the TOM content showed some variation but values exceeding 1.0% were rare. No significant differences could be detected (Fig. 1b, ANOVA, $p = 0.117$).

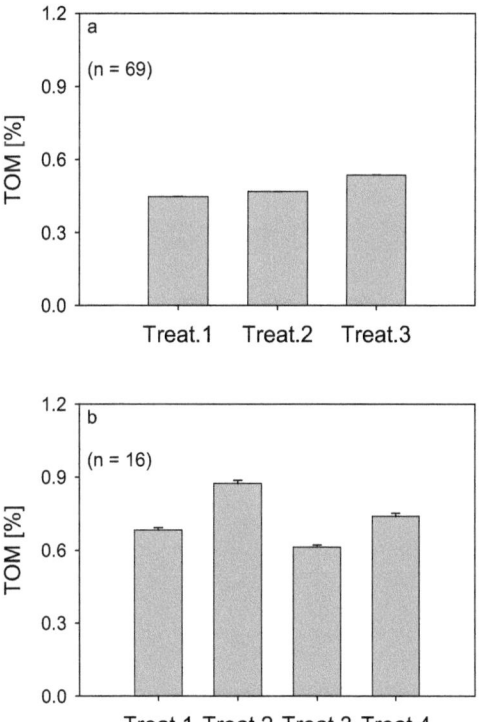

Fig. 1: Total organic matter content (means + SE) in [%] of the sediments during both experiments with brown shrimp (*C. crangon*): a) TOM content of the treatments (Treat.1 – Treat.3) in the first experiment; b) TOM content of the treatments (Treat.1 – Treat. 4) in the second experiment.

3.5 Survival of *C. crangon*

After finishing both experiments survival rates of the shrimps were calculated by using the initial and final numbers of *C. crangon*. Fig. 2 displays the survival rates calculated for both experiments. A large fraction of the animals used for the first experiment died during the experiment. Highest survival rate for the individual treatments during the first experiment was 33%. The total calculated survival rate of the first experiment, using all animals recaptured from the growth units, was 21%. The highest survival rate for the individual treatments during the second experiment was 93%. For the second experiment the total survival rate, using the recaptured animals from all growth units, was 62%. The total survival rate, using all recaptured animals from the entire system (tanks plus sump) was 81%.

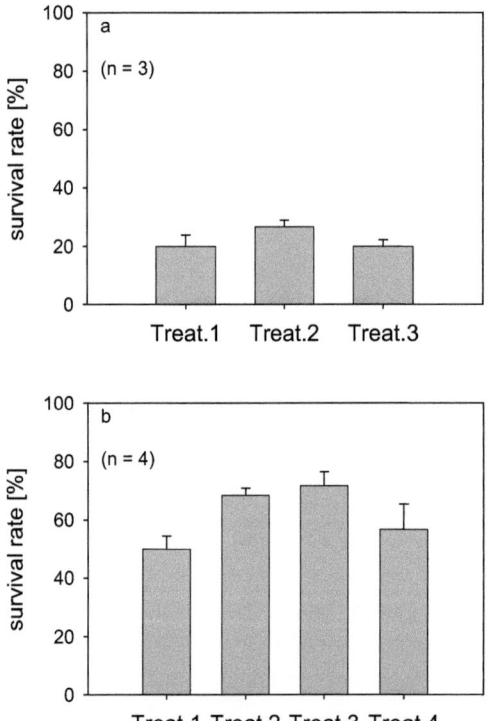

Fig. 2: Survival rates of brown shrimps during both experiments: a) survival obtained from the three treatments (Treat.1 – Treat.3) of the first experiment; b) survival of the four treatments (Treat.1 – Treat. 4) from the second experiment. Bars represent means + SE.

3.6 Growth of *C. crangon*

Growth was calculated as the weight difference between initial and final wet weight. Fig. 3 represents the average initial and final weights of *C. crangon* of both experiments. Animals used for the first experiment were older and consequently heavier compared to animals used for the second experiment.

After the first experiment negative weight changes could be detected whereas, for the second experiment positive weight gains for most of the treatments could be detected. Treatment 2 was the only exception where no weight change appeared.

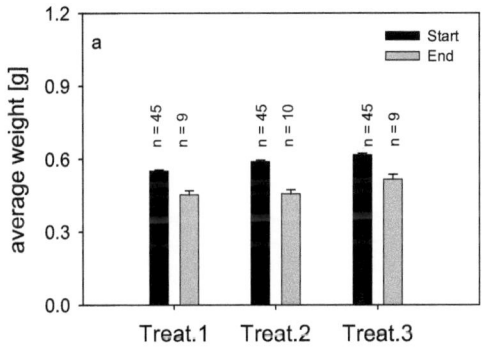

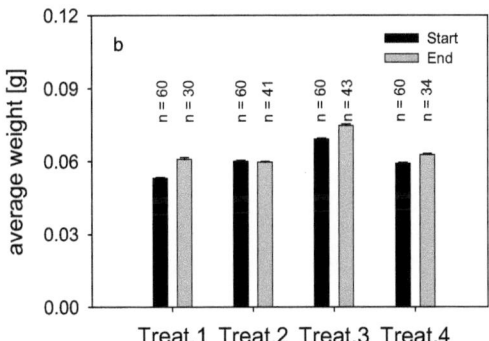

Fig. 3: Average weight of *C. crangon* at the start and the end of the experiments: a) initial and final weight for the first experiment; b) initial and final weight for the second experiment. Bars represent means + SE.

The average weight of the shrimps recorded during the experiments was used to calculate the specific growth rates µ in $[d^{-1}]$ according to Jørgensen (1990). Tab. 6 and Fig. 3 indicate for the first experiment negative growth rates and for the second experiment positive growth rates or at least no weight losses for treatment 2 of the second experiment.

Tab. 6: Specific growth rates µ $[d^{-1}]$ obtained for *C. crangon* during both experiments.

	Experiment 1	Experiment 2
Treatment 1	-0.010	0.006
Treatment 2	-0.013	0.000
Treatment 3	-0.009	0.003
Treatment 4		0.003

3.7 Biochemical composition of *C. crangon*

The biochemical analyses of the shrimps included the water, total organic matter and energy content. Furthermore, carbon and nitrogen concentrations of animal tissues were recorded.

Fig. 4 presents the water content of both experiments. Initial values of both experiments were comparable, showing no difference for small (experiment 2; Fig. 4b) and large (experiment 1; Fig. 4a) individuals. After experimental application lower water contents were found for larger individuals (< 80%, experiment 1) compared to smaller individuals (> 80%, experiment 2).

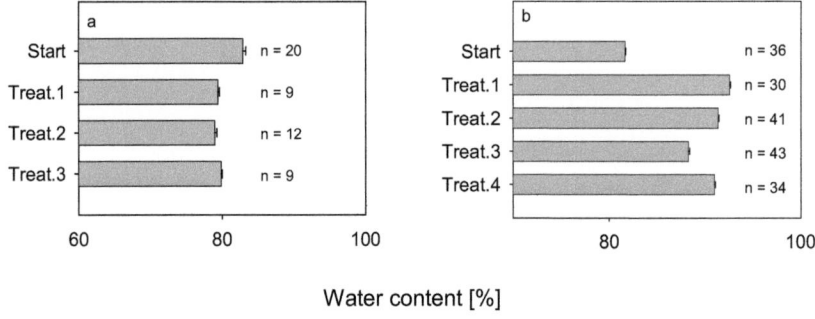

Water content [%]

Fig. 4: Initial and final water contents of brown shrimps: a) water contents obtained from the first experiment; b) water contents obtained from the second experiment. Bars represent means + SE.

Variations within treatments during the first experiment were detectable. For treatment 2 this difference was significant (Fig. 4a; ANOVA, p = 0.028) showing lower water contents compared to the start group. With increasing amounts of food during the first experiment the water content was increasing. For the second experiment the initial water content of 82% was the lowest recorded during this experiment. Treatments 1, 2 and 4 showed similar water contents ranging around 91%. Treatment 3 of the second experiment showed with 88% a lower water content than the other treatments. After the second experiment differences concerning the water content between the start group and all treatments were significant (Fig. 4b; ANOVA, p < 0.001). Differences occurring among treatments were also significant between treatment 3 and all others (ANOVA, p < 0.001).

The initial and final total organic matter content of *C. crangon* in both experiments is shown in Fig. 5. Some variations during the first experiment are detectable showing highest values around 71% at the start and decreasing contents during the experiment, reaching lowest values of approx. 68% for the starvation group. TOM contents of the start group and treatment 3 were almost identical whereas the TOM content of treatment 1 and 2 were lower than the start group but no significant difference could be detected (ANOVA, p = 0.284). For the second experiment these variations between treatments increased, showing the highest TOM content (around 71%) also at the start of the experiment. Treatment 1, fed with the lowest amount of solids, showed the lowest TOM content followed by the starvation group (treatment 4), treatment 2, fed with the high amount of solids and treatment 3, fed with fish feed, respectively. Except treatment 3, in which the commercial fish feed was fed, treatments showed significant differences compared to the start group (Fig. 5b; ANOVA, p<0.001). Treatment 1 showed also significant differences compared to treatment 3 (Fig. 5b; ANOVA, p = 0.001).

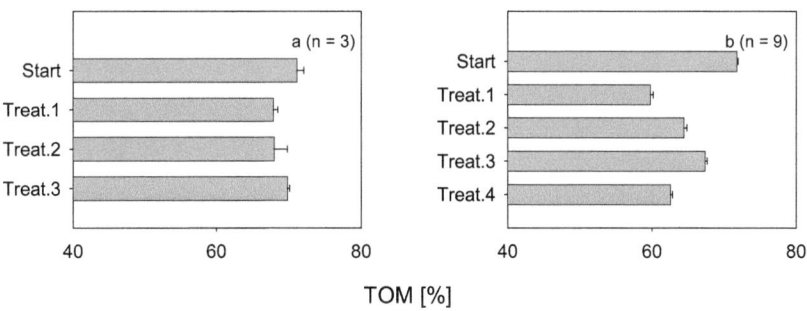

Fig. 5: Initial and final TOM content of european brown shrimps: a) TOM contents obtained from the first experiment; b) TOM contents obtained from the second experiment. Bars represent means + SE.

Energy contents of *C. crangon*, measured as gross energy in [kJ g^{-1} DW] is presented in Fig. 6 for the start groups and all treatments applied during the experiments. Start values obtained from both experiments were comparable showing values above

16 kJ g^{-1} DW. For the first experiment energy content decreased for all treatments and this differences were significant (Fig. 6a; ANOVA, p < 0.001). During the second experiment the energy contents increased, reaching values above 20 kJ g^{-1} DW. All treatments increased their energy content but for none of the treatments these differences were significant (Fig. 6b; ANOVA, p = 0.080).

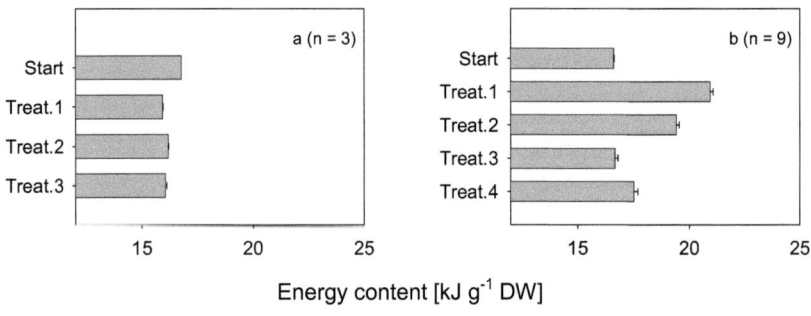

Fig. 6: Initial and final energy content [kJ g^{-1} DW] of *C. crangon*: a) energy contents obtained from the first experiment; b) energy contents obtained from the second experiment. Bars represent means + SE.

Carbon and nitrogen concentrations of the shrimps are presented in Figs. 7 and 8. Clear tendencies were observed in both experiments indicating higher carbon and nitrogen concentrations associated with increasing amounts of food and improved food quality.

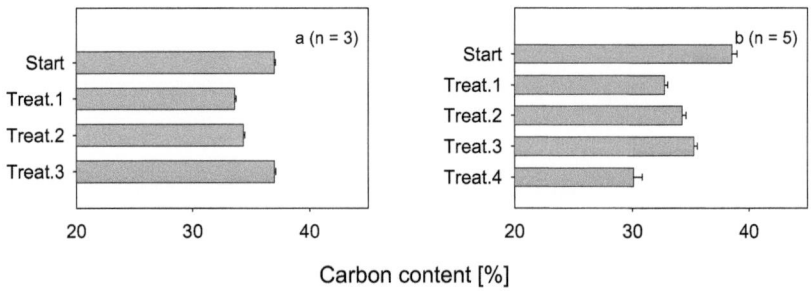

Fig. 7: Initial and final carbon concentrations of the shrimps: a) carbon concentrations obtained from the first experiment; b) carbon contents obtained from the second experiment. Bars represent means + SE.

During the first experiment carbon concentrations obtained from the start group and treatment 3 were identical. The other two treatments showed a significant decrease of carbon (Fig. 7a; ANOVA, p < 0.001). For the second experiment the situation was similar. With decreasing amounts of food and decreasing quality of food average carbon concentration decreased. The highest carbon concentrations were observed for treatment 3, fed commercial fish feed and the lowest carbon concentration was found for treatment 4, the starvation group. Differences between the start group and treatments 1 and 4 were significant (Fig. 7b; ANOVA, p < 0.001 for treatment 4 and p = 0.014 for treatment 1).

Nitrogen concentrations of shrimps showed similar patterns. For the first experiment no differences between the start group and treatment 3 could be detected but for treatments 1 and 2 compared to the start group differences were significant (Fig. 8a; ANOVA, p < 0.001). For the second experiment differences between the start group and treatments 1 and 4 were significant (Fig. 8b; ANOVA, p = 0.005 for treatment 4 and p = 0.017 for treatment 1).

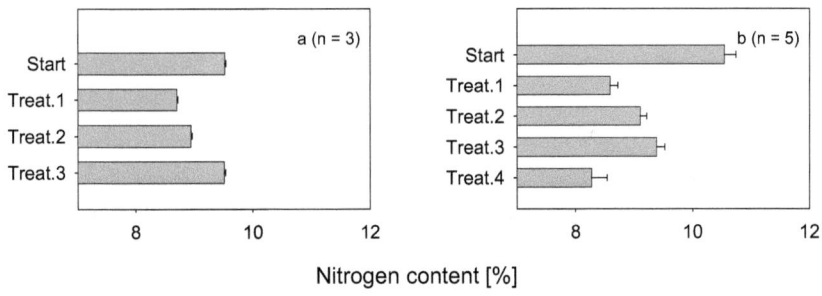

Fig. 8: Initial and final nitrogen concentrations of brown shrimps: a) nitrogen concentrations obtained from the first experiment; b) nitrogen contents from the second experiment. Bars represent means + SE.

4. Discussion

Environmental conditions including temperature, salinity, pH, concentrations of dissolved oxygen and inorganic nutrients during experiment 1 were not suitable for the culture of *C. crangon*. Improvements during the second experiment could be observed. Salinity was too low during the first experiment and was consequently adapted to natural conditions for the second experiment. According to Sommer (1998) crustaceans like *Gammarus duebenii* and *Carcinus maenas* are hypertonic regulators that means they have to control their water flow in response to different ion concentrations. This may also be necessary for *C. crangon* and this osmotic requirement can have caused stress to the animals during the first experiment.

pH values for natural seawater range from 8.1 to 8.3 at CO_2-equillibrium with the atmosphere (Sommer 1998). Kater *et al.* (2006) reported strong reductions in LC_{50} values for total ammonium concentrations as soon as pH was exceeding values of 8.4. Magallon-Barajas *et al.* (2006) recommended for optimal survival of post-larvae from *Litopenaeus vannamei* during transport pH values not higher than 8. The elevated pH values during experiment 1 probably adds to the stress levels of the organisms. During the second experiment pH values were in the range of natural seawater.

Oxygen consumption of 0.75 ml h^{-1} $tank^{-1}$ for both experiments can be calculated according to Regnault (1981). During both experiments the oxygen saturation was constantly above 5 ml L^{-1}, respectively. Therefore no negative effects due to an undersupply of oxygen were expected.

The dissolved inorganic nutrients during both experiments were low and within save limits for all nutrients except the TAN concentrations in the first experiment. Nitrate and phosphate concentrations never exceeded critical values. According to Masser *et al.* (1999) nitrite concentrations as low as 0.5 mg L^{-1} already caused stress to catfish but values up to 5.0 mg L^{-1} caused no stress for *Tilapia*. It was assumed that the observed nitrite concentrations had no effect on the shrimps during the experiments. TAN concentrations during the first experiment reached values of up to 0.15 mg L^{-1} and unionised ammonia was 0.07 mg L^{-1}. These concentrations were in the range of concentrations described by Masser *et al.* (1999) responsible for slow growth and tissue damage for several fish species. During the second experiment concentrations of unionised ammonia was around 0.01 mg L^{-1} and we assumed that these concentrations had no effect on the shrimps during the second experiment.

Water content of the applied solids was in general agreement with natural food sources for *C. crangon* such as plant and animal detritus or small prey (e.g. worms). Values for the TOM content varied between the three different food sources. The TOM content which consists of proteins, carbohydrates and fat, is responsible for the energy delivered by the food. Due to digestion by fish the energy contents of solid waste in the form of faecal material is reduced compared to fish feed. These values might be still sufficient for the metabolic requirements of organisms low in the food chain such as *C. crangon*. Consequently, in relation to the TOM content the concentrations of total carbon and nitrogen showed similar patterns. Total carbon and nitrogen concentrations are comparable or even exceeding natural systems (pers. communication H. Thetmeyer). Thus, it is assumed that fish feed and solid waste in the forms presented should have met the metabolic requirements of the cultured animals.

Average TOM content of the sediments obtained from all treatments of both experiments was low compared to natural sediments. A tendency associated with the applied treatments was detectable, with increasing amounts of food the TOM content of the sediment increased. With increasing TOM contents of the sediment it seems doubtful that *C. crangon* is able to consume the complete food. Further investigations should try to answer the question whether the applied feeding rate was oversized or the food composition was not adequate.

Survival rates of shrimps during the first experiment were low. These low survival rates might be explained by combining the factors inducing stress to the animals. Factors such as high temperatures, short acclimatisation periods, low salinities, elevated pH values and increased concentrations of unionised ammonia cause reduced survival of invertebrate animals. Due to improved environmental conditions survival could be increased and total survival exceeded 80%. Comparing the results of both experiments it seems obvious that due to improved environmental conditions the survival improved and reached values described by Regnault (1976), where mortality ranged between 40 and 66% during experimental conditions.

During the first experiment all treatments showed a reduction of average weight. This could be due to the food quality. Protein requirements of shrimps decrease with age according to Regnault (1978), who reported a decrease in required optimum protein diet contents for shrimps from initially 70 to 10% during an increase of body length from 13 to 30 mm. During the first experiment animals reached average weights of 500 to 600 mg and it can be assumed that the diet composition was sufficient to meet the metabolic requirements. Another explanation for the poor growth performance of *C. crangon* under the applied conditions could be the high stress levels induced by unfavourable abiotic parameters and unionised ammonia. As a consequence the animals reduced their feed intake and therefore were forced to use their own reserves for metabolism. Supporting this assumption is the fact, that although different amounts of food were applied to the treatments, the average weight reduction for treatments 1 (no food applied) and 3 (high amount of food applied) were similar. For treatment 2 (medium amount of food applied) this value was even higher. During the second experiment the average weight kept constant or increased, even for treatment 4, the starvation group. Due to high variations no significant differences for growth could be detected but specific growth rates µ reaching values up to 0.006 d^{-1} were recorded for the second experiment. The highest wet weight gain was 0.57 mg which was approx. ten times lower compared to Regnault (1976), where daily wet weight gains of 5.8 mg were achieved under similar conditions. No clear tendency between the amount of food and specific growth rates µ or the quality of food (solids or fish feed) and growth could be detected. This leads to the assumption that both food sources are equally suitable for the culture of *C. crangon* and that the lower amount of food was sufficient to achieve growth of shrimps but this growth was very low. The composition of the diet might have a strong impact on the growth. For the second experiment the average weight of the shrimps was low and therefore protein requirements were high according to Regnault (1978). Ingested food remains only a short time (up to a few hours) in the guts of most crustaceans (Dall and Moriarty 1983), which is not likely to be a sufficient time for bacterial fermentation to contribute substantially to the digestion. Bacteria might function as suppliers of vitamins or essential amino acids but the essential requirements of components for synthesis and energy have already to be available in the food.

No differences between the amounts of food applied during the treatments and the resulting specific growth rates µ or the quality of food and specific growth rates µ could be detected. However for the biochemical composition of the shrimps fed with different amounts and quality of food such differences were observed.

TOM content of the shrimps showed some variation for both experiments. Regnault (1981) reported an average TOM content for *C. crangon* of 62% (dry weight) for normally fed individuals and up to 78% for well fed individuals. The values obtained during both experiments were within this range of normal fed animals. For the first experiment TOM contents of the start group and treatment 3, the group fed with the high amounts of solids, were almost identical indicating that the amount of food supplied during the experiment was sufficient to maintain the biochemical composition of total organic matter of the animals over a period of 21 days. The TOM content of treatment 1 and 2 were lower than the start group but no significant difference could be detected. During the second experiment differences between the start group and the treatments were more distinct. All treatments lost TOM during the course of the experiment but were above concentrations of 42% described by Regnault (1981) for starving shrimps. TOM content decreased more with lower amounts of food and decreasing food quality. The energy content of *C. crangon* also changed significantly during the experiments. Energy contents obtained from the first experiment showed similar values for the treatments but lower values compared to the start group. This is in agreement with the decreasing TOM content observed from the biochemical composition of the shrimps. In the second experiment the situation changed completely. Energy contents of cultured individuals increased. An explanation of this unexpected result could be the low amounts of available sample material which could have caused errors during analyses. Another explanation for the observed energy increase could be the weather conditions prior to the collection of the shrimps. Water temperatures were low (ice cover was still occasionally occuring) and little food might be available for the shrimps causing starvation of the animals prior to the experiment. But this is in contrast to the carbon and nitrogen concentrations obtained from the shrimps. During the course of the experiments carbon and nitrogen were decreasing indicating an undersupply with these essential nutrients.

Using the carbon and nitrogen concentrations recorded during both experiments and calculating the ratio between carbon and nitrogen loss during both experiments, it appears that for the starvation groups this ratio is close to 1. Consequently we assumed that both elements were equally used for metabolic processes. The situation observed for the treatments where food was applied changed. During the first experiment the carbon loss was higher than the nitrogen loss. During the second experiment the ratio changed and the nitrogen loss was higher than the carbon loss. An explanation for this could be the different experimental temperatures. During higher temperatures metabolic rates are elevated and therefore the amount of required energy is increased. The high amount of energy will be supplied by the metabolism of fat, which delivers more energy compared to proteins; fat also contains a high number of carbon atoms. For the second experiment the temperatures were lower and as a result the amount of energy required by the animals was reduced and could be supplied by the use of proteins which deliver less energy compared to fat. An alternative explanation may be the low nitrogen content supplied with food during the second experiment. Nitrogen consumption of the animals was higher than the supply, whereas the carbon supply was still sufficient, leading to a slower decrease of carbon by the metabolism.

Cannibalism might be another factor strongly influencing the results of the presented experiments. Cannibalism is well known and described for *C. crangon* (Plagmann 1939; Tiews 1954; Regnault 1976; Del Norte-Campos and Temming 1994). Water treatment applied during both experiment reduced turbidity in the water and enables the shrimps to see and prey on each other. It can be assumed that cannibalism had an impact on all treatments especially considering the low quality of the applied food.

5. Conclusions

Farming *C. crangon* as secondary organisms in multitrophic integrated aquaculture, fed with solids seems possible. It appeared to be important to adjust the environmental parameters such as temperature, salinity, pH values and concentrations of dissolved inorganic nutrients also to the requirements of the additionally farmed organisms. Integrating the cultivation of *C. crangon* into the culture of cold water adapted species, such as trout or cod, could lead to reduced amounts of solid waste from these fish cultures.

Survival of brown shrimps during the experiments could be increased above 80 %. Positive growth rates under the applied conditions were achieved and environmental parameters were suitable for the culture of *C. crangon*.

Nevertheless, the biochemical composition of the analysed shrimps indicated that further research is necessary before the integration of *C. crangon* into aquaculture will be successful. This research should focus on the appropriate amounts of food required for maintenance and growth of the shrimps. Growth rates were low compared to natural conditions. Future scientific questions should include size related feeding regimes. Another focus should be the composition of the administered food. Solid waste from aquaculture contains reduced concentrations of essential nitrogen. Sublimation of solid waste through the culture of heterotrophic bacteria, algae and/or meiobenthic organisms and thus, developing short artificial food chains adapted to the nutritional needs of *C. crangon* should be investigated. A third field of investigation should be the impacts of stocking densities on intraspecific behaviour such as induced stress by elevated abundances or cannibalism. The last area of research, which should be focussed on is reproduction under cultural conditions. To establish a self sustaining brood stock is a crucial step to successfully achieve sustainability for shrimp culture.

Chapter 7

General Conclusions

Bischoff A.A.

This study was carried out to investigate the potential of a secondary detritivorous reactor to mitigate the environmental impacts of recirculating aquaculture systems. This final chapter combines the results from this study and relevant questions include:

- Is it possible to reduce the solid waste load from recirculating aquaculture systems by the cultivation of detritivorous organisms?

- What are the benefits of producing secondary organisms?

- Which steps towards sustainability can be achieved?

- Which criteria need to be fulfilled for a successful integration of detritivorous organisms into aquaculture?

7.1 Is it possible to reduce the solid waste load from recirculating aquaculture systems by the cultivation of detritivorous organisms?

It could be demonstrated that it is possible to achieve a reduction in solid waste from a recirculating aquaculture system.

The positive growth rates obtained from both organisms, utilised during the experiments (*Nereis diversicolor* and *Crangon crangon*), demonstrated that even with a limited food quality (solid waste excreted from fish) growth and a biomass increase of the cultured taxa is achievable. Under the nutritional conditions applied in this study, specific growth rates (SGR) of *N. diversicolor* as high as 0.025 d^{-1} were recorded.

In the experiments with the shrimps (*C. crangon*) the culture conditions could be optimized. Survival of *C. crangon* under cultural conditions exceeded that under natural conditions and positive growth was achieved (SGR = 0.006 d^{-1}).

The concentrations of total organic matter in the sediment can not be used as a reliable indicator for the consumption of solid waste by detritivorous organisms. However, an additional indicator, fatty acid composition, was also investigated and proved to be a more reliable indicator. By comparing fatty acid profiles of fish feed, excreted solid waste, sediments, biofilms and worm tissues, it could be shown that *N. diversicolor* fed on settling materials, including uneaten fish feed and solid waste.

7.2 What are the benefits of producing secondary organisms?

7.2.1 Increased consumption of supplied nutrients

The consumption of otherwise wasted nutrients was increased due to the culture of detritivorous organisms. Applying the results obtained during this research, the nutrient balance of the bioreactor for detritivorous organisms can be calculated. The worms had to consume between 60 and 100 % of the settled solid waste, to meet their carbon and nitrogen demands.

7.2.2 Reduction of water exchange of recirculating aquaculture systems

Due to the design of the integrated recirculating system no water was lost during the treatment for solid waste removal. The solid waste, produced in the self cleaning fish tanks, was transferred via the water flow to the worm bioreactor, which also acts as a settling tank. During both experiments performed within the integrated system, the daily water exchange rate in the recirculated aquaculture system was below 1% of the total system volume.

7.2.3 Economical diversification of aquaculture endeavours

Cultivation of *N. diversicolor* or *C. crangon* yielded additional harvestable biomass with high market potentials.

Assuming a successful harvest of the worms and comparing the costs of fish feed used during the cultivation period of *N. diversicolor* and the potential profit from selling the worms, a financial benefit exceeding three times the cost of the feed could have been achieved.

These calculations are based on the achieved experimental data, which do not account for initial investment costs, amortisation and labour and to also not take into account the holding, processing and marketing costs. However such data may be helpful to improve calculations when considering scale up.

7.3 Which steps towards sustainability can be achieved?

The seed supply is one major step towards sustainability. In this study a self-sustaining broodstock of *N. diversicolor* was established in an integrated recirculating system. A sufficient worm supply could be met with offspring from the recirculating system, and no further wild caught individuals were required.

By introducing a detritivorous reactor as secondary step into the system it was possible to reduce the water exchange rates of the recirculated aquaculture system to 0.8% d^{-1} of the system volume. These values were far below values, postulated for closed recirculating systems (> 10% d^{-1} system volume). Thus consumption of the valuable water resources could be reduced. Due to the nutrient removal from the effluent waters by the production of secondary biomass, a reduced environmental impact of aquaculture could be achieved through diminished nutrient discharge in effluent waters.

7.4 Which criteria need to be fulfilled to integrate successfully detritivorous organisms into aquaculture?

In general only a few important requirements have to be fulfilled:

1. A continuous production of faecal material from the target cultivation species such as fish or crustaceans.

2. A well designed recirculating aquaculture system, which transfers the faecal material with no additional effort to the secondary species.

3. A suitable secondary species, which meets its growth requirements from the nutritional composition of the faecal material in a suitable bioreactor.

4. Survival of secondary aquaculture organisms must be high.

5. A self-sustaining broodstock of the secondary organism should be established.

6. For an implementation into conventional recirculating aquaculture systems, it is necessary to sell the cultured organism and therefore gain financial revenues from the cultivation of such an organism.

References

A

Acinas S. G., Anton J. and Rodriguez-Valera F. (1999). Diversity of free-living and attached bacteria in offshore western Mediterranean waters as depicted by analysis of genes encoding 16S rRNA. Appl. Environ. Microbiol., Vol. 65, no. 2: 514 – 522

Ackefors H. and Enell M. (1994). The release of nutrients and organic matter from aquaculture systems in Nordic countries. J. Appl. Ichthyol. 10: 225 – 241

Ahn O., Petrell R. J., and Harrison P. J. (1998). Ammonium and nitrate uptake by *Laminaria saccharina* and *Nereocystis luetkeana* originating from a salmon sea cage farm. J. Appl. Phycol. 10: 333 – 340

Alterman D. J., Rosenthal H., Smith P., Stewart J. and Weston D. (1994). Chemicals used in aquaculture. ICES Coorp. Res. Report 202, 1 – 100 pp

Altmann D., Stief P., Amann R., DeBeer D. and Schramm A. (2003). In situ distribution and activity of nitrifying bacteria in freshwater sediment. Environ. Microbiol. 5, 798-803

Altmann D., Stief P., Amann R., and DeBeer D. (2004). Distribution and activity of nitrifying bacteria in natural stream sediment versus laboratory sediment microcosms. Aquat. Microb. Ecol. 36, 73-81

Andrade L. R., Farina M. and Amado Filho G. M. (2004). Effects of copper on *Enteromorpha flexuosa* (Chlorophyta) in vitro. Ecotoxicol. Environ. Saf. 58: 117 – 125

Andrew M. I. and Frank L. (2004). Integrated aquaculture system for nutrient reduction in agricultural wastewater: potential and challenges. Bull. Fish. Res. Agency 1: 143 – 152

Asgard T., Austreng E., Holmefjord I., and Hillestad M. (1999). Resource efficiency in the production of various species. *In*: Svennevig N., Reinertsen N., New H., and New M. (eds). Sustainable aquaculture: food for the future? A. A. Balkema, Rotterdam, Holland.: 171-183

B

Baeyens W., Meulemann C., Muhaya B. and Leermakers M. (1998). Behaviour and speciation of mercury in the Scheldt estuary (water, sediments and benthic organisms). Hydrobiologia 366: 63 – 79

Baird M. E. and Middleton J. H. (2004). On relating physical limits to the carbon: nitrogen ratio of unicellular algae and benthic plants. J. Mar. Syst. 49: 169 – 175

Barak Y. and van Rijn J. (2000). Biological phosphate removal in a prototype recirculation aquaculture treatment system. Aquacult. Eng. 22: 121-136

Batista F. M., Fidalgo e Costa P., Matias D., Joaquim S., Massapina C., Passos A. M., Pousao Ferreira P. and Cancela da Fonseca L. (2003). Preliminary results on the growth and survival of the polachaete *Nereis diversicolor* (O. F. Müller, 1776), when fed with faeces from the carpet sheel clam Ruditapes decussatus (L., 1758). Bol. Inst. Esp. Oceanogr. 19 (1 – 4): 443 – 446

Belgrano A. (2005). Aquatic food webs' ecology: old and new challenges. In Belgrano A., Scharler U. M., Dunne J. and Ulamovlcz R. E. (eds.) Aquatic food webs. Oxford University Press, Oxford

Bell J. G., Tocher D. R., MacDonald F. M. and Sargent J. R. (1994). Effects of diets rich in linoleic (18:2n – 6) and α-linolenic (18:3n – 3) acids on the growth, lipid class and fatty acid compositions and eicosanoid production in juvenile turbot (*Scophthalmus maximus* L.). Fish Physiol. Biochem. 13: 105 – 118

Bell J. G. and Sargent J. R. (2003). Arachidonic acid in aquaculture feeds: current status and future opportunities. Aquaculture 218: 491 – 499

Berthet B., Mouneyrac C., Amiard J. C., Amiard-Triquet C., Berthelot Y., Le Hen A., Mastain O., Rainbow P. S. and Smith B. D. (2003). Accumulation and soluble binding of cadmium, copper, and zinc in the polychaete *Hediste diversicolor* from coastal sites with different trace metal bioavailabilities. Arch. Environ. Contam. Toxicol. 45: 468 – 478

Bischoff A. A. (2003). Growth and mortality of the polychaete *Nereis diversicolor* under experimental rearing conditions. M.Sc.thesis, Institute of Marine Research & Department of Animal Sciences, Chairgroup of Fish Culture and Fisheries, Christian Albrechts University Kiel, Germany/Wageningen University, The Netherlands; 103pp

Bordin G., Cordeiro Raposo F., McCourt J. and Rodriguez A. R. (1994). Identification of metallothionein-like metal binding proteins in the marine bivalve *Macoma baltica*. Life Sci. 317: 1057 – 1064

Boujard T., Gelineau A. Coves D., Corraze G. Dutto G., Gasset E. and Kaushik S. (2004). Regulation of feed intake, growth, nutrient and energy utilisation in

European sea bass (*Dicentrarchus labrax*) fed high fat diets. Aquaculture 231: 529 – 545

Bovendeur J. (1989). Fixed-biofilm reactors applied to waste water treatment and aquacultural water recirculating systems. Thesis, Agricultural University of Wageningen, The Netherlands

Bray W. A. and Lawrence A. L. (1992). Reproduction of *Penaeus* species in captivity. In Fast A. W. and Lester J. L. (eds.) Marine shrimp culture: principle and practices. Elsevier, Amsterdam, p 93 – 170

Brügmann L. (ed.). Meeresverunreinigungen. Akademie-Verlag, Berlin, pp. 294 (1993)

Bryan G. W. (1974). Adaptation of an estuarine polychaete to sediments containing high concentrations of heavy metals. In Vernberg F.J. and Vernberg W.B. (eds.) Effects on Pollution on the Physiological Ecology of Estuarine and Coastal Water Organisms, Academic Press, NY, USA

Budge S. M., Iverson S. J., Bowen W. D. and Ackman R. G. (2002). Among and within-species variability in fatty acid signatures of marine fish and invertebrates on the Scotian Shelf, Georges Bank, and southern Gulf of St. Lawrence. Can. J. Fish. Aquat. Sci. 59: 886 – 898

Bullock G., Herman R., Heinen J., Weber A. and Hawkins J. (1994). Observations on the occurrence of bacterial gill disease and amoeba gill infestation in rainbow trout cultured in a water recirculation system.' J. Aquat. Anim. Health 6: 310 – 317

Buschmann A. H., Troell M., Kautsky N., and Kautsky L. (1996). Integrated tank cultivation of salmonids and *Gracilaria chilensis* (Rhodophyta). Hydrobiologica 326/327: 75-82

Buschmann A. H., Hernandez-Gonzalez M. C., Astudillo C., de la Fuente L., Gutierrez A. and Aroca G. (2005). Seaweed cultuivation, product development and integrated aquaculture studies in Chile. World Aquacult. 36: 51 – 53

Buschmann A. H., Riquelme V. A., Hernandez-Gonzalez M. C., Varela D., Jimenez J. E., Henriquez L. A., Vergara P. A., Guinez R. and Filun L. (2006). A review of the impacts of salmonid farming on marine coastal ecosystems in the southeast Pacific. ICES J. Mar. Sci. 63: 1338 – 1345

C

Caballero M. J., Izquierdo M. S., Kjorsvik E., Fernandez A. J. And Rosenlund G. (2004). Histological alterations in the liver of sea bream, Sparus aurata L., caused by short- or long-term feeding with vegetable oils. Recovery of normal morphology after feeding fish oil as the sole lipid source. J. Fish Dis. 27: 531 – 541

Chamberlain G. and Rosenthal H. (1995). Aquaculture in the next century: Opportunities for growth challenges of sustainability. World Aquaculture 26: 21 – 25

Chambers, M. R. and Milne, H. (1975). Life cycle and production of *Nereis diversicolor* O. F. Mueller in the Ythan Estuary, Scotland. Estuar. Coast. Shelf Sci. 3: 133 – 144

Cho C. Y. and Bureau D. P. (2001). A review of diet formulation strategies and feeding systems to reduce excretory and feed wastes in aquaculture. Aquacult. Res. 32: 349 – 360

Chopin T., Sharp G., Belyea E., Semple R., and Jones D. (1999a). Open water aquaculture of the red algae *Chondrus crispus* in Prince Edward Island, Canada. Hydrobiologica 398/399: 417 – 425

Chopin T. Yarish C., Wilkes R., Belyea E., Lu S. and Mathieson A (1999b). Developing Porphyra/salmon integrated aquaculture for bioremediation and diversification of the aquaculture industry. J. Appl. Phycol. 11: 463 – 472

Chopin T., Buschman A. H., Halling C., Troell M., Kautsky N., Neori A., Kraemer G. P., Zertuche-Gonzales J. A., Yarish C., and Neefus C. (2001). Integrating seaweeds into marine aquaculture systems: A key toward sustainability. J. Phycol. 37: 975 – 986

Chopin T. and Bastarache S. (2004). Mariculture in Canada: Fisnfish, shellfish and seaweed. World Aquacult. 35: 37 – 41

Chopin T., Sawhney M., Shea R., Belyea E., Bastarache S., Armstrong W., Boyne-Travis S., Robinson S., Ridler N., Sewuster J. and Szemerda M. (2006). Kelp (*Laminaria saccharina*) aquaculture as the inorganic extractive component of an Integrated Multi-Trophic Aquaculture (IMTA) system with salmon (*Salmo salar*) and mussels (*Mytilus edulis*) in the Bay of Fundy, New Brunswick, Canada. AQUA 2006 Linking Tradition & Technology, Highest Quality for the consumer. Special publication

Colloca F. and Cerasi S. (2005). Cultured Aquatic Species Information Programme - *Sparus aurata*. FAO Inland Water Resources and Aquaculture Service (FIRI). FAO - FIGIS

Collos Y. and Slawyk G. (1980). Uptake and assimilation by marine phytoplankton. *In:* Falkowski P.G. Uptake and assimilation by marine phytoplankton. Plenum Press, New York, pp. 195-211

Costa-Pierce B. (1996). Environmental impacts of nutrients discharged from aquaculture: towards the evolution of sustainable, ecological aquaculture systems, In Baird D. J., Beveridge M. C. M., Kelly L. A. and Muir J. F. (eds.) Aquaculture and water resource management. p. 81-113. Blackwell Science, Oxford, UK

Costa-Pierce B. (ed.) Ecological aquaculture: the evolution of the blue revolution. Blackwell Publishing, Oxford, UK, 2002. 382 pp

Costa P. F. E. (1999). Reproduction and growth in captivity of the polychaete *Nereis diversicolor* O.F. Müller, 1776, using two different kinds of sediment: Preliminary assays. Bol. Inst. Esp. Oceanogr. 15: 351 – 355

Costa P. F. E., Narciso L. and Cancela da Fonseca L. (2000). Growth, survival and fatty acid profile of *Nereis diversicolor* (O.F. Mueller, 1776) fed on six different diets. Bull. Mar. Sci. 67: 337 – 343

Cripps S. J. and Bergheim A. (2000). Solids management and removal for intensive land-based aquaculture production systems. Aquacult. Eng. 22: 33-56

Curds C. R. (1982). The ecology and role of protozoa in aerobic sewage treatment processing. Annu. Rev. Microbiol. 36: 27 – 46

D

Daims H., Nielsen J. L., Nielsen P. H., Schleifer K. H. and Wagner M. (2001). In situ characterization of Nitrospira-like nitrite-oxidizing bacteria active in wastewater treatment plants. Appl. Environ. Microbiol. 67, 5273-5284

Dales R. P. (1950). The reproduction and larval development of *Nereis diversicolor* O. F. Mueller. J. Mar. Biol. Assoc. U.K. 29: 321 – 360

Dall W. and Moriarty D. J. W. (1983). Functional aspects of nutrition and digestion. In Mantel LH (ed.) The biology of crustacea Vol. 5. Academic Press, New York

Dalsgaard T., Canfield D. E., Petersen J., Thamdrup B. and Acuna-Gonsaléz J. (2003a). N_2 production by the anammox reaction in the anoxic water column of Golfo Dulce, Costa Rica. Nature 422: 606 – 608

Dalsgaard J., St John M., Kattner G., Müller-Navarra D. and Hagen W. (2003b). Fatty acid trophic markers in the pelagic marine environment. Adv. Mar. Biol. 46: 225 – 340

Davey J. T. (1994). The architecture of the burrow of *Nereis diversicolor* and its quantification in relation to sediment–water exchange. J. Exp. Mar. Biol. Ecol. 179, 115– 129

Davis D. A. (2001). Best management practices for feed and feeding practices. Aquaculture 2001: Book of Abstracts, p. 166

Dean, A. C. R. (1957). The adaptation of bacterial cultures during the lag phase in media containing new substrates or antibacterial agents. Proc. R. Soc. London, Ser. B 147: 247 – 257

Del Norte-Campos A. G. C. and Temming A. (1994). Daily activity, feeding and rations in gobies and brown shrimp in the northern Wadden Sea. Mar. Ecol. Prog. Ser. 115:41 – 53

Dhainaut A., Raveillon D., Mberi M., Porchethennere E. and Demuynck S. (1989). Purification of an antibacterial protein in the celomic fluid of *Nereis diversicolor* (Annelida, Polychaeta) - Similitude with a cadmium-binding protein. Comp. Biochem. Physiol. Pt. C 94: 555 – 560

Dollhopf S. L., Hyen J. H., Smith A. C., Adams H. J., O'Brien S. and Kostka J. E. (2005). Quantification of Ammonia-Oxidizing Bacteria and Factors Controlling Nitrification in Salt Marsh Sediments. Appl. Environ. Microbiol. 71: 240 – 246

Dosdat A., Servivas F., Metailler R., Huelvan C. and Desbruyeres E. (1996). Comparison of nitrogenous losses in five teleost fish species. Aquaculture 141: 107 – 127

E

Edwards P. (1998). A systems approach for the promotion of integrated aquaculture. Aquacult. Econ. Manage. 2: 1 – 12

Edwards P. and Pullin R.S.V. (1990). Wastewater-fed aquaculture. ICLARM No. 684, Bangkok, Thailand, 296pp

Elder D. L., Fowler S. W. and Polikarpov G. G. (1979). Remobilization of sediment-associated PCBS by the worm *Nereis diversicolor*. Bull. Environ. Contam. Toxicol. 21: 48 – 452

Enell M. and Loef J. (1983). Environmental impacts of aquaculture-sedimentation and nutrient loadings from fish cage culture farming. Vatten 39: 364 – 375

Erler D. V., Pollard P. C., Burke M. and Knibb W. (2000). Biological remediation of aquaculture waste: a combined finfish, artificial substrate treatment system.

Proceedings of the National Workshop of Wastewater Treatment and Integrated Aquaculture, pp 93 – 107

Erler D., Pollard P., Duncan P. and Knibb W. (2004). Treatment of shrimp farm effluent with omnivorous finfish and artificial substrates. Aquacult. Res. 35: 816 – 827

F

Fantin A. M. B. and Franchini A. (1990). Ultrastructural changes in the ganglia of *Viviparus ater* following experimental lead intoxication. J. Invertebr. Pathol. 56: 387 – 394

FAO (2006a). Fisheries Technical Paper 500. State of World Aquaculture. Editorial Group FAO Information Division. 134pp

FAO (2006b). FISH INFOnetwork Market Report on Fishmeal (November 2006) http://www.eurofish.dk/indexSub.php?id=3378
(Ref type: Electronic Citation; date 03.01.2007)

Fast A. W. and Menasveta P. (2000). Some recent issues and innovations in marine shrimp pond culture. Rev. Fish. Sci. 8: 151 – 233

Fauchald K. and Jumars P. A. (1979). The diet of worms: a study of polychaete feeding guilds. Oceanogr. Mar. Biol. Ann. Rev. 17: 193 – 284

Ferns P. N. and Anderson J. I. (1997). Lead in the diet and body tissues of dunlins, *Calidris alpine*, from the Bristol Channel, UK. Environ. Pollut. 96: 35 – 42

Fontenot Q. C., Isely J. J. and Tomasso J. R. (1998). Acute toxicity of ammonia and nitrite to shortnose sturgeon fingerlings. Progressive Fish Culturist 60: 315 – 318

Franco-Nava M. A., Blancheton J. P., Deviller G. and Le-Gall J.Y. (2004). Particulate matter dynamics and transformations in a recirculating aquaculture system: application of stable isotope tracers in seabass rearing. Aquacult. Engin. 31: 135 – 155

Frangipane G., Ghirardini A. V., Collarini F., Zaggia L., Pesce A. and Tagliapietre D. (2005). Heavy metals in *Hediste diversicolor* (polychaeta : nereididae) and salt marsh sediments from the lagoon of Venice (Italy). Chem. Ecol. 21: 441 – 454

Fraser A. J., Sargent J. R., Gamble J. C. and Seaton D. D. (1989). Formation and transfer of fatty acids in an enclosed marine food chain comprising phytoplankton, zooplankton and herring (*Clupea harengus*) larvae. Mar. Chem. 27: 1 – 18

Fuller S. A., Henne J. R., Carmichael G. J. and Tomasso J. R. (2003). Toxicity of ammonia and nitrite to the Gila trout. N. Am. J. Aquacult. 65: 162 – 164

Furuita H., Unuma T., Nomura K., Tanaka H., Okuzawa K., Sugita T. and Yamamotot T. (2006). Lipid and fatty acid composition of eggs producing larvae with high survival rate in the Japanese eel. J. Fish Biol. 69: 1178 – 1189

G

Gill I. and Valivety R. (1997). Polyunsaturated fatty acids, part I: occurrence, biological activities and applications. Trends Biotechnol. 15: 401 – 409.

Graeve M., Hagen W. and Kattner G. (1994). Herbivorous or omnivorous? On the significance of lipid compositions as trophic markers in Antarctic copepods. Deep Sea Res. 26: 281 – 347

Grasshoff K., Kremling K. and Erhardt M. (eds.) Methods of seawater analysis – Third, completely revised and extended edition. Wiley-VCH, Weinheim, Germany (1999)

Grossmann S. and Reichardt W. (1991). Impact of *Arenicola marine* on bacteria in intertidal sediments. Mar. Ecol. Prog. Ser. 77: 85 – 93

Guillen J. L., Endo M., Turnbull J. F., Kawatsu H, Richards R. H. and Aoki T. (1993). Depressed growth rate and damage of the cartilage of red sea bream, associated with exposure to ammonia. Bull. Jap. Soc. Sci. Fish. 59: 1231 – 1234

Gyllenhammar A. and Håkanson L. (2005). Environmental consequence analyses of fish farm emissions related to different scales and exemplified by data from the Baltic – a review. Mar. Environ. Res. 60: 211 – 243

H

Hansen K. and Kristensen E. (1997). Impact of macrofaunal recolonization on benthic metabolism and nutrient fluxes in a shallow marine sediment previously overgrown with macroalgal flats. Estuar. Coast. Shelf Sci. 45: 613 – 628

Hall P. O. J., Kollberg S. and Samuelsson M. O. (1992). Chemical fluxes and mass balance in a marine fish cage farm. IV. Nitrogen. Mar. Ecol. Prog. Ser. 89: 81 – 91

Hardy R. W. and Tacon A. G. J. (2002). Fish Meal: Historical Use, Production Trends and Future Outlook for Sustainable Supplies. In Stickney R.R. and McVey J.P. (eds.) 'Responsible Marine Aquaculture', CABI Publishing, Wallingford, Oxon, UK

Hargreaves J. A. (1998). Nitrogen biogeochemistry of aquaculture ponds. Aquaculture 166: 181 – 212

Harris S. A. and Probyn T. (1996). Nitrogen excretion and absorption efficiencies of white steenbras, *Lithognathus lithognathus* Cuvier (Sparidae), under experimental culture conditions. Aquacult. Res. 27: 43 – 56

Hartmann-Schröder G. (ed.) Annelida, Borstenwürmer, Polychaeta. Gustav Fischer Verlag, Jena (1996)

Harvey R. W. and Luoma S. N. (1994). The role of bacterial exopolymer and suspended bacteria in the nutrition of the deposit-feeding clam *Macoma balthica*. J. Mar. Res. 42: 957 – 968

Hastings N., Agaba M., Tocher D. R., Leaver M. J., Dick J. R., Sargent J. R. and Teale A. J. (2001). A vertebrate fatty acid desaturase with $\Delta 5$ and $\Delta 6$ activities. PNAS 98: 14304 – 14309

Hateley J. G., Grant A. and Jones N. V. (1989). Heavy metal tolerance in estuarine populations of *Nereis diversicolor*. In Ryland J. S. and Tyler P. A. (eds.) 23. European Marine Biology Symp., Swansea (UK), 5 – 9 Sep 1988

Heise S., Koch C., Krost P. and Piker L. (1996). Die Einflüsse der Fischerei und Aquakultur auf die marine Umwelt. Texte zum Umweltforschungsplan des Bundesministeriums für Umwelt, Naturschutz und Reaktorsicherheit – Wasserwirtschaft: 46/96, 136 pp

Heilskov A. C. and Holmer, M. (2001). Effects of benthic fauna on organic matter mineralization in fish-farm sediments: importance of size and abundance. ICES J. Mar. Sci. 58: 427 – 434

Heissenberger A., Lepperd G. G. and Herndl G. J. (1996). Ultrastructure of marine snow. II Microbiological considerations. Mar. Ecol. Prog. Ser. 135: 299 – 308

Hershberger W. K. (2002). Genetic Changes in Marine Aquaculture Species and the Potential for Impacts on Natural Populations. In Stickney R.R. and McVey J.P. (eds.) 'Responsible Marine Aquaculture', CABI Publishing, Wallingford, Oxon, UK

Hillestad M. and Johnson F. (1994). High-energy/low-protein diets for Atlantic salmon: Effects on growth, nutrient retention and slaughter quality. Aquaculture 124: 109 – 116

Hovanec T. A., Taylor L. T., Blakis A., Delong E. F. (1998). *Nitrospira*-Like Bacteria Associated with Nitrite Oxidation in Freshwater Aquaria. Appl. Environ. Microbiol. 64: 258 – 264

Hölting B. (1996). Hydrogeologie, Vol. Enke-Verlag, Stuttgart

Hursthouse A. S., Matthews J. M., Figures J. E., Iqbal-Zahid P., Davies I. M. and Vaughan D. H. (2003). Chromium in intertidal sediments of the Clyde, UK: Potential for remobilisation and bioaccumulation. Environ. Geochem. Health. 25: 171 – 203

I

Ibarz A., Blasco J., Beltran M., Gallardo M. A., Sanchez J., Sala R. and Fernandez-Borras J. (2005). Cold-induced alterations on proximate composition and fatty acid profiles of several tissues in gilthead sea bream (*Sparus aurata*). Aquaculture 249: 477 – 486

Ibeas C., Rodriguez C., Badia P., Cejas J. R., Santamaria F. J. and Lorenzo A. (2000). Efficacy of dietary methyl esters of $n-3$ HUFA vs. triacylglycerols of $n-3$ HUFA by gilthead sea bream (*Sparus aurata* L.) juveniles. Aquaculture 190: 273 – 287

ICES Mariculture Committee (2003). Report of the working group on environmental interactions of mariculture (WGEIM). ICES CM 2003/F:04 1-114

Imsland A. K., Foss A., Sparboe L. O. and Sigurdsson S. (2006). The effect of temperature and fish size on growth and feed efficiency ratio of juvenile spotted wolffish *Anarhichas minor*. J. Fish. Biol. 68: 1107 – 1122

Islam M., Sarker M., Yamamoto T., Wahab M. and Tanaka M. (2004). Water and sediment quality, partial mass budget and effluent N loading in coastal brackish water shrimp farms in Bangladesh. Mar. Pollut. Bull. 48: 471 – 485

Iverson S.J. (1993). Milk secretion in marine mammals in relation to foraging: can milk fatty acids predict diet? Symp. Zool. Soc. Lond. 66: 263 – 291

J

Jiang J., Ke S. and Ding L. (2004). Phosphorous states, distribution and environmental impacts in the sediments of cage aquaculture. J. Zhejiang Ocean Univ. 23: 311 – 314

Jørgensen C. B. (1990). Bivalve filter feeding: hydrodynamics, bioenergetics, physiology and ecology. Olsen & Olsen, Fredensborg

K

Kater B. J., Dubbeldam M. and Postma J. F. (2006). Ammonium toxicity at high pH in a marine bioassay using *Corophium volutator*. Arch. Environ. Contam. Toxicol. 51: 347 – 351

Kim H. G. (1997). Recent harmful algal blooms mitigating strategies in Korea. Ocean Research 19: 185 – 192

Kim J. D, Kaushik S. J. and Breque J. (1998). Nitrogen and phosphorous utilisation in rainbow trout (Oncorhynchus mykiss) fed diets with and without fish meal. Aquat. Living Resour. 11: 261 – 268

Kirsch P. E., Iverson S. J., Bowen W. D., Kerr S. R. and Ackman R. G. (1998). Dietary effects on the fatty acid signature of whole Atlantic cod (*Gadus morhua*). Can. J. Fish. Aquat. Sci. 55: 1378 – 1386

Kirsch P. E., Iverson S. J. and Bowen W. D. (2000). Effect of diet on body composition on blubber fatty acids in captive harp seals (*Phoca groenlandica*). Physiol. Biochem. Zool. 73: 45 – 59

Kolbe K. (1995). Heavy metals in sediments, zoobenthos and vegetation of the eulittoral of the Weser estuary. Ber. Forschungsstelle Kueste 40: 39 – 54

Koenneke M., Bernhard A. E., De la Torre J. R., Walker C. B., Waterbury J. B. and Stahl D. A. (2005). Isolation of an autotrophic ammonia-oxidizing marine archaeon. Nature 437: 543 – 546

Koven W., Barr Y., Lutzky S., Ben-Atia I., Weiss R., Harel M., Behrens P. and Tandler A. (2001). The effect of dietary arachidonic acid (20:4 n-6) on growth, survival and resistance to resistance to handling stress in gilthead seabream (*Sparus aurata*) larvae. Aquaculture 193: 107 – 122

Kristensen E. (1984). Effect of natural concentrations on nutrient exchange between a polychaete burrow in estuarine sediment and the overlying water. J. Exp. Mar. Biol. Ecol. 75, 170 – 190

Kristensen E. (2000). Organic matter diagenesis at the oxic/anoxic interface in coastal marine sediments, with emphasis on the role of burrowing animals. Hydrobiologia 426: 1 – 24

Krom M. D., Porter C., and Gordin H. (1985). Nutrient budget of a marine fish pond in Eilat, Israel. Aquaculture 51: 65 – 80

Krom M. D. and Neori A. (1989). A total nutrient budget for an experimental intensive fishpond with circularly moving seawater. Aquaculture 83: 345 – 358

Krom M. D., Ellner S., van Rijn J., and Neori A. (1995). Nitrogen and phosphorous cycling and transformations in a prototype 'non-polluting' integrated mariculture system, Eilat, Israel. Mar. Ecol. Prog. Ser. 118: 25 – 36

Kube N. (2006). The integration of microalgae photobioreactors in recirculation systems for low water discharge mariculture. PhD-thesis at the Leibniz-Institute of Marine Science, Kiel, Germany, pp 170

Kube N. and Rosenthal H. (2006). Ozonation and foam fractionation used for the removal of bacteria and particles in a marine recirculation system for microalgae cultivation. In Kube N. (ed.) The integration of microalgae photobioreactors in recirculation systems for low water discharge mariculture. PhD-thesis at the Leibniz-Institute of Marine Science, Kiel, Germany. (2006)

Kumar M., Ingerson T. and Lewis R. (2000). Wastewater treatment and integrated aquaculture: South Australian initiatives and international collaboration. Proceedings of the National Workshop on Wastewater Treatment and Integrated Aquaculture. Pp. 14 – 18

Kuwae T., Hosokawa Y. (1999). Determination of Abundance and Biovolume of Bacteria in Sediments by Dual Staining with 4´,6-Diamidino-2-Phenylindole and Acridine Orange: Relationship to Dispersion Treatment and Sediment Characteristics. Appl. Environ. Microbiol. 65: 3407 – 3412

L

Lachner A. (1972). Ammoniak als Stoffwechselprodukt beim Fisch unter besonderer Berücksichtigung der Ammoniak-Stickstoff-Ausscheidungen bei Streßsituationen. In Probleme der Ernährung und Haltung von Süßwasserfischen im Intensivbetrieb. Münchner Beiträge zur Abwasser-, Fischerei- und Flussbiologie 23: 23 – 41

Lee W-Y and Wang W-X (2001). Metal accumulation in the green macroalgae *Ulva fasciata*: effects of nitrate, ammonium and phosphate. Sci. Total Environ. 278: 11 – 22

Lehninger A. L., Nelson D. L. and Cox M. M. (eds.) Principles of Biochemistry, 2nd Edition, Worth Publishers, N.Y. 1993

Lemarie G., Dosdat A., Coves D., Dutto G., Gasset E. and Person-Le Ruyet J. (2004). Effect of chronic ammonia exposure on growth of European seabass (*Dicentrarchus labrax*) juveniles. Aquaculture 229: 479 – 491

Levasseur M., Thompson P. A. and Harrison P. J. (1993). Physiological acclimation of marine phytoplankton to different nitrogen sources. J. Phycol. 29: 587 – 595

Linares F. and Henderson R. J. (1991). Incorporation of ^{14}C-labelled polyunsaturated fatty acids by juvenile turbot, Scophthalmus maximus (l.) in vivo. J. Fish Biol. 38: 335 – 347

Lin C. K., Ruamthaveesub P. and Wanuchsoontorn P. (1993). Interated culture of the green mussel (Perna viridis) in wastewater from an intensive shrimp pond: Concept and practice. World Aquacult. 24: 68 – 73

Llobet-Brossa E., Rosseló-Mora R. and Amann R. (1998). Microbial community composition of wadden sea sediments as revealed by fluorescence in situ hybridization. Appl. Environ. Microbiol. 65: 3407 – 3412

Losordo T. M., Masser M. P., and Rakocy J. E. (1999). Recirculating aquaculture tank production systems: A review of component options. SRAC Publication 453

Lotz J. M. (1997). Viruses, biosecurity and specific pathogen-free stocks in shrimp aquaculture. World Journal of Microbiology and Biotechnology 13: 405 – 413

Loy A., Horn M. and Wagner M. (2003). probeBase - an online resource for rRNA-targeted oligonucleotide probes. Nucleic Acids Res. 31, 514 – 516

Lucas F. and Bertru G. (1997). Bacteriolysis in the gut of Nereis diversicolor (O. F. Mueller) and effect of the diet. J. Exp. Mar. Biol. Ecol. 215: 235 – 245

Lucas J. S. and Southgate P. C. (eds.) Aquaculture – Farming aquatic animals and plants. Blackwell Publishing Ltd, Oxford, UK (2003)

Lucas F. S., Bertru G. and Höfle M.G. (2003). Characterization of free-living and attached bacteria in sediments colonized by Hediste diversicolor. Aquat. Microb. Ecol. 32: 165 – 174

Luis O. J. and Passos A. M. (1995). Seasonal changes in lipid content and composition of the polychaete Nereis (Hediste) diversicolor. Comp. Biochem. Physiol. 111: 579 – 586

Lupatsch I., Kissil G. W., Sklan D. and Pfeffer E. (2001). Effects of varying dietary protein and energy supply on growth, body composition and protein utilization in Gilthead seabream (Sparus aurata L.). Aquacult. Nutr. 7: 71 – 80

Lupatsch I. and Kissil G. W. (1998). Predicting aquaculture waste from Gilthead Seabream (*Sparus aurata*) culture using a nutritional approach. Aquat. Living Ressour. 11: 265 – 268

Lupatsch I., Kissil G. W., and Sklan D. (2003). Comparison of energy protein efficiency among three fish species gilthead sea bream (*Sparus aurata*), European sea bass (*Dicentrarchus labrax*) and white grouper (*Epinephelus aeneus*): energy expenditure for protein and lipid deposition. Aquaculture 225: 175 – 189

M

Madigan M. T., Martinko J. M., and Parker J. (eds) Brock Mikrobiologie. Spektrum Akademischer Verlag Berlin Heidelberg, (2001)

Magallon-Barajas F. J., Servin Villegas R., Portillo Clark G. and Lopez Moreno B. (2006). *Litopenaeus vannamei* (Boone) post-larval survival related to age, temperature, pH and ammonium concentration. Aquacult. Res. 37: 492 – 499

Masser M. P., Racocy J. and Losordo T. M. (1999). Recirculating aquaculture tank production systems – management of recirculating systems. SRAC publications No. 452

Matos J., Costa S., Rodrigues A., Pereira R. and Sousa-Pinto I. (2006). Experimental integrated aquaculture of fish and red seaweeds in Northern Portugal. Aquaculture 252: 31 – 42

McHatton S. C., Barry J. P., Jannasch H. W. and Nelson D. C. (1996). High Nitrate Concentrations in Vacuolate, Autotrophic Marine Beggiatoa spp.. Appl. Environ. Microbiol. 62: 954 – 958

McVey J. P., Stickney R. R., Yarish C. and Chopin T. (2002). Aquatic Polyculture and Balanced Ecosystem Management: New Paradigms for Seafood Production. In Stickney R. R. and McVey J. P. (eds.) Responsible Marine Aquaculture, CABI Publishing, Wallingford, Oxon, UK

Metaxa E., Deviller G., Pagand P., Alliaume C., Casellas C., and Blancheton J. P. (2006). High rate algal pond treatment for water reuse in a marine fish recirculation system: Water purification and fish health. Aquaculture 252: 92 – 101

Mettam C. (1979). Seasonal changes in populations of *Nereis diversicolor* O. F. Müller from the Severn estuary, U. K. In Cyclic Phenomena in Marine Plants and Animals (Edited by Naylor E. and Hartnoll R. G), pp. 123 – 130. Pergamon Press, Oxford

Mettman C., Santhanam V. and Harvard M. S. C. (1982). The oogenic cycle of *Nereis diversicolor* under natural conditions. J. Mar. Biol. Assoc. U. K. 62: 637 – 645

Milanese M., Chelossi E., Manconi R., Sara A., Sidri M. and Pronzato R. (2003). The marine sponge *Chondrilla nucula* Schmidt, 1862 as an elective candidate for bioremedation in integrated aquaculture. Biomol. Eng. 20: 363 – 368

Mobarry, B. K., Wagner, M., Urbain, V., Rittmann, B. E. and Stahl D. A. (1996). Phylogenetic probes for analyzing abundance and spatial organization of nitrifying bacteria. Appl. Environ. Microbiol. 62: 2156 – 2162

Montagna, P. A. (1984) In situ measurement of meiobenthic grazing rates on sediment bacteria and edaphic diatoms. Mar. Ecol. Prog. Ser. 18: 119 – 130

Moriarty D. J. W. and Hyaward A. C. (1982). Ultastructure of bacteria and the proportion of gram negative bacteria in marine sediments. Microb. Ecol. 8: 1 - 14

N

Neori A., Cohen I. and Gordin H. (1991). *Ulva lactuca* biofilters for marine fish-pond effluents. II. Growth rate, yield and C:N ratio. Bot. Mar. 34: 483-489

Neori A., Shpigel M. and Ben-Ezra, D. (2000). A sustainable integrated system for culture of fish, seaweed and abalone. Aquaculture 186: 279-291

Neori A., Chopin T., Troell M., Buschmann A. H., Kraemer G. P., Halling C., Shpigel M. and Yarish C. (2004). Integrated aquaculture: rationale, evolution and state of the art emphasizing seaweed biofiltration in modern mariculture. Aquaculture 231: 361 – 391

Newkirk G. (1996). Sustainable coastal production systems: a model for integrating aquaculture and fisheries under community management. Ocean Coast. Manag. 32: 69 – 83

Nielsen A. M., N., Eriksen T., Iversen J. J. L. and Riisgard H. U. (1995). Feeding, Growth and Respiration in the Polychaetes *Nereis diversicolor* (Facultative Filter-Feeder) and *N. virens* (Omnivorous) - a Comparative-Study. Mar. Ecol. Prog. Ser. 125: 149 – 158

O

Ogle J. T. and Lotz J. M. (1998). Preliminary design of a closed, biosecure shrimp grow out system. Proceedings of the US Marine Shrimp Farming Program Biosecurity Workshop

Olive P. J. M. (1999). Polychaete aquaculture and polychaete science: a mutual synergism. Hydobiologia 402: 175 – 183

Olivier M., Desrosiers G., Caron A., Retiere C. and Caillou A. (1995). Behavioral responses of *Nereis diversicolor* (O.F. Mueller) and *Nereis virens* (Sars) (Polychaeta) to food stimuli - Use of specific organic matter (algae and halophytes). Can. J. Zool. 73: 2307 – 2317

Ottolenghi F., SilvestriC., Giordano P., Lovatelli A., and New M. B. (2004). Capture-based aquaculture. The fattening of eels, groupers, tunas and yellowtails. Rome, FAO, 308 pp

Owen J. M., Adron J. W., Middleton C. and Cowey C. B. (1975). Elongation and desaturation of dietary fatty acids in turbot *Scophthalmus maximus* and rainbow trout *Salmo gairdneri*. Lipids 10: 528 – 531

P

Pearson T. H. and Rosenberg R. (1978). Macrobenthic succession in relation to organic enrichment and pollution of the marine environment. Oceanogr. mar. Biol. A. Rev. 16: 229 – 311

Pernthaler J., Glöckner F.O., Schönhuber W. and Amann R. (2001). Fluorescence in situ hybridization (FISH) with rRNA-targeted oligonucleotide probes. In: Paul H (ed.) Methods in Mircobiology, Vol 30. Academic Press, San Diego

Peres H. and Oliva-Teles A. (2006). Effect of the dietary essential to non-essential amino acid ratio on growth, feed utilization and nitrogen metabolism of European sea bass (*Dicentrarhcus labrax*). Aquaculture 256: 395 – 402

Petrell R. J. and Alie S. Y. (1996). Integrated aquaculture of salmonids and seaweeds in open systems. Hydrobiologica 326/327: 67 – 73

Plagmann J. (1939). Ernährungsbiologie der Garnele (*Crangon vulgaris* Fabr.). Helgoländer wiss. Meeresunters. 2 (1): 113 - 162

Plante C. J., Jumars P. A. and Baross A. (1989). Rapid bacterial growth in the hindgut of a marine deposit feeder. Microb. Ecol. 18: 29 – 44

Plante C. J. and Mayer L. M. (1994). Distribution and efficiency of bacteriolysis in the gut of *Arenicola marina* and three additional deposit feeders. Mar. Ecol. Prog. Ser. 109: 183 – 194

Porter K. G. and Feig Y. S. (1980). The use of DAPI for identifying and counting aquatic microflora. Limnol. Oceanogr. 25: 943 – 948

Porter C. B., Krom M. D. and Gordin H. (1986). The effects of water quality on the growth of Sparus aurata in marine fish ponds. Aquaculture 59: 299 – 315

Pousao P., Machado M. and Cancela da Fonseca L. (1995). Marine pond culture in southern Portugal: Present status and future perspectives. Cah. Options Mediterr. Seminar of the CIHEAM Network on Technology of Aquaculture in the Mediterranean (TECAM), Nicosia (Cyprus), 14. – 17. Jun. 1995

Preston N. P., Crocos P. J., Keys S. J., Coman G. J. and Koenig R. (2004). Comparative growth of selected and non-selected Kuruma shrimp *Penaeus (Marsupenaeus) japonicus* in commercial farm ponds; implications for broodstock production. Aquaculture 231: 73 – 82

R

Rabalais N. N. (2002). Nitrogen in aquatic ecosystems. Ambio 31: 102 – 112

Ramaiah N. (2005). Role of heterotrophic bacteria in marine ecological processes in Marine Microbiology: Facets & Opportunities. Published by the National Institute of Oceanography, Dona Paula, Goa

Readman J. W., Kwong L. L. W., Grondin D., Bartocci J., Villeneuve J. P. and Mee L. D. (1993). Environ. Sci. Technol. 27: 1940 – 1942

Regnault M. (1976). Influence of the substrate on the mortality and growth of the shrimp *Crangon crangon* (L.) in rearing. Cah. Biol. Mar. 17: 347 – 357

Regnault M. (1978). Quantitative protein requirement of the sand shrimp *C. crangon* and effect of a compound diet on growth. Oceanis 4: 73 – 81

Regnault M. (1981). Respiration and ammonia excretion of the shrimp *Crangon crangon* L.: metabolic response to prolonged starvation. J. Comp. Physiol. 141: 902 – 912

Reichardt W. (1988). Impact of bioturbation by Arenicola marina on microbiological parameters in intertidal sediments. Mar. Ecol. Prog. Ser. 44: 149 – 158

Rheinheimer G., Hegemann W. and Sekoulov R. J. (eds.) Stickstoffkreislauf im Wasser: Stickstoffumsetzung in natürlichen Gewässern, in der

Abwasserreinigung und Wasserversorgung. Oldenburg Verlag, München, 1988

Ringwood A. H. and Keppler C. J. (1998). Seed clam growth: An alternative sediment bioassay developed during EMAP in the Carolinian Province. Environ. Monit. Assess. 511: 247 – 257

Riisgard H. U., Poulsen L. and Larsen P. S. (1996). Phytoplankton reduction in near-bottom water caused by filter-feeding *Nereis diversicolor* - implications for worm growth and population grazing impact. Mar. Ecol. Prog. Ser. 141, 47 – 54

Risgaard-Petersen N., Nicolaisen M. H., Revsbech N. P. and Lomstein B. A. (2004). Competition between Ammonia-Oxidizing Bacteria and Benthic Microalgae. Appl. Environ. Microbiol. 70: 5528 – 5537

Rodriguez C., Acosta C., Badia P., Cejas C. R., Santamaria F. J. and Lorenzo A. (2004). Assessment of lipid and essential fatty acids requirements of black sea bream (*Spondyliosoma cantharus* L.) by comparison of lipid composition in muscle and liver of wild and captive adult fish. Comp. Biochem. Physiol. B: Biochem. Mol. Biol. 139: 619 – 629

Russell N. J. and Nichols D. S. (1999). Polyunsaturated fatty acids in marine bacteria – a dogma rewritten. Microbiology 145: 767 – 779

Ryther J.H. (1983). The evolution of integrated aquaculture systems. J. World Maricult. Soc. 14: 473 – 484

S

Saiz-Salinas J. I. and Frances-Zubillaga G. (1997). Enhanced growth in juvenile *Nereis diversicolor* after its exposure to anaerobic polluted sediments. Mar. Pollut. Bull. 34: 437 – 442

Sargent J. R., Parkes R. J., Mueller-Harvey I. and Henderson R. J. (1987). Lipid biomarkers in marine ecology. In Sleigh M.A. (ed.) Microbes in the sea, Wiley and Sons, New York, pp: 119 – 138

Sargent J. R., Bell J. G., Bell M. V., Henderson R. J. and Tocher D. R. (1995). Requirement criteria for essential fatty acids. J. Appl. Ichthyol. 11: 183 – 198

Sargent J., McEvoy L., Estevez A., Bell G., Bell M., Henderson J. and Tocher D. (1999). Lipid nutrition of marine fish during early development: current status and future directions. Aquaculture 179: 217 – 229

Sawyer T. K. and Davis E. E. (1989). Protozoans, fungi, and bacteria as indicators of coastal contamination related to ocean waste disposal practices. J. Shellfish Res. 8, pp 481

Scaps P. (2002). A review of the biology, ecology and potential use of the common ragworm *Hediste diversicolor* (O. F. Muller) (Annelida: Polychaeta). Hydrobiologia 470: 203 – 218

Schneider O., Sereti V., Eding E. H., and Verreth J. A. J. (2005). Analysis of nutrient flows in integrated intensive aquaculture systems. Aquacult. Eng. 32: 379 – 401

Schneider O. (2006). Fish waste management by conversion into heterotrophic bacteria biomass. PhD-thesis at the Fisheries department at Wageningen University, The Netherlands. 160 pp

Schwoeberl J. (1999). Einführung in die Limnologie. Gustav Fischer Verlag, Stuttgart

Septier F., Demuynck S., Thomas P. and Dhainaut-Courtois N. (1991). Bioconcentration of heavy metals in an estuarine annelida polychaete: *Nereis diversicolor*. Oceanol. Acta 11 Int. Samp. Sur l'Environment des Mers Epicontinentales, Lille (France) ; 20 – 22 Mar 1990

Sherr E. B., Sherr B. F. (2002). Significance of predation by protists in aquatic microbial food webs. Antonie van Leeuwenhoek 81, 293-308

Shucksmith R., Hinz H., Bergmann M. and Kaiser M. J. (2006). Evaluation of habitat use by adult plaice (*Pleuronectes platessa* L.) using underwater video survey techniques. J. Sea Res. 56: 317 – 328

Sijtsma L. and de Swaaf M. E. (2004). Biotechnological production and application of the ω-3 polyunsaturated fatty acid docosahexaenoic cid. Appl. Microbial Biotechnol. 64: 146 – 153

Sommer U. (ed.) Biologische Meereskunde. Springer Verlag, Berlin (1998)

Spiek E. and Bock E. (1998). Taxonomische, physiologische und ökologische Vielfalt nitrifizierender Bakterien. Biospektrum 4: 25 – 31

Stickney R. R. (2002). Issues associated with Non-indigenous Species in Marine Aquaculture. In Stickney R.R. and McVey J.P. (eds.) 'Responsible Marine Aquaculture', CABI Publishing, Wallingford, Oxon, UK

Støttrup J. G. and McEvoy L. A. (eds.) Live feeds in marine aquaculture. Blackwell publishing, 336pp, (2002)

Sudaryono A., Hoxey M. J., Kallis S. G. and Evans L. H. (1995). Investigation of alternative protein sources in practical diets for juvenile shrimps, *Penaeus monodon*. Aquaculture 134: 313 – 323

Summerfelt S. T. (2002). An integrated approach to aquaculture waste management in flowing water systems. Proceedings of the 2nd International Conference on Recirculating Aquaculture: 87-97

Suvapepun S. (1994). Environmental impacts of mariculture. Chulalongkorn Univ., Bangkok, Thailand: pp. 25 – 29

Sweerts J. P. R. A., De Beer D., Nielsen L. P., Verdouw H., Van den Heuvel J. C., Cohen Y., and Cappenberg T. E. (1990). Denitrification by sulphur oxidizing Beggiatoa spp. mats on freshwater sediments. Nature 344: 762 – 763

Syakti A. D., Mazzella N., Torre F., Acquaviva M., Gilewicz M., Guiliano M., Betrand J. C. and Doumenq P. (2006). Influence of growth phase on the phospholipidic fatty acid composition of two marine bacterial strains in pure and mixed cultures. Res. Microbiol. 157: 479 – 486

T

Tatrai I. (1986). Influence of temperature, rate of feeding and body weight on nitrogen metabolism of bream, *Abramis brama* L.. Comp. Biochem. Physiol. A: Physiol.: 83A: 543 – 547

Terlizzi A., Scuderi D., Faimali M., Minganti V. and Geraci S. (1997). Imposex in *Hexaplex trucnulus* and*Stramonita haemastoma* (Gatsropoda, Muricida): first records for the Italian coasts. Biol. Mar. Mediterr. 4: 496 – 499

Tiews K. (1954). Die biologischen Grundlagen der Büsumer Garnelenfischerei. Ber. Dt. Komm. Meeresforschung 13: 270 – 282

Tilak K. S., Lakshmi S. J. and Susan T. A. (2002). The toxicity of ammonia, nitrite and nitrate to the fish, *Catla catla* (Hamilton). J. Environ. Biol. 23: 147 – 149

Tisdell C. A. (1995). Economics, the environment and sustainable aquaculture: extending the discussion. The University of Queensland, Department of Economics, Discussion Papers 180: 0 – 19

Tocher D. R. and Ghioni C. (1999). Fatty acid metabolism in marine fish: Low activity of fatty acyl $\Delta 5$ desaturation in gilthead sea bream (*Sparus aurata*) cells. Lipids 34: 43 – 440

Tocher D. R., Fonseca-Madrigal J., Dick J. R., Ng W.-K., Bell J. G. And Campell P. J. (2004). Effects of water temperature and diets containing palm oil on fatty acid desaturation and oxidation in hepatocytes and intestinal enterocytes of rainbow trout (*Oncorhynchus mykiss*). Comp. Biochem. Physiol. Pt. B 137: 49 – 63

Tovar A., Moreno C., Manuel-Vez M.P. and Garcia-Vargas M. (2000). Environmental impacts of intensive aquaculture in marine waters. Water. Res. 34: 334 – 342

Troell M., Halling C., Neori A., Chopin T., Buschman A. H., Kautsky N. and Yarish C. (2003). Integrated mariculture: asking the right questions. Aquaculture 226: 69-90

U

UK Marine – Special Areas of Conservation Project www.ukmarinesac.org.uk/activities/bait-collection/bc9_1.htm
(Ref type: Electronic Citation; date 28.08.2006)

V

Vandermeulen H. and Gordin H. (1990). Ammonium uptake using *Ulva* (Chlorophyta) in intensive fishpond systems: mass culture and treatment of effluent. J. Appl. Phycol. 2: 363 – 374

Vedel A. and Riisgard H. U. (1993). Filter-Feeding in the Polychaete *Nereis diversicolor* - Growth and Bioenergetics. Mar. Ecol. Prog. Ser. 100: 145 – 152

Verhagen F. J. M. and Laanbroek H. J. (1992). Effects of grazing by flagellates on competition for ammonium between nitrifying and heterotrophic bacteria in chemostats. Appl. Environ. Microbiol. 58: 1962 – 1969

Vijayan K. K., Stalin Raj V., Balasubramanian C. P., Alavandi S. V., Thillai Sekhar V. and Santiago T. C. (2005). Polychaete worms – a vector for white spot syndrome virus (WSSV). Dis. Aquat. Org. 63: 107 – 111

von Elert E. and Stampfl P. (2000). Food quality for *Eudiaptomus gracilis*: the importance of particular highly unsaturated fatty acids. Freshw. Biol. 45: 189 – 200

von Halem O. (2006). Numerische Modellierung der Nährstoffdynamik einer integrierten Aquakultur-Kreislaufanlage. MSc.Thesis, University of Oldenburg, Germany, 80 pp

W

Wagner M., Rath G., Koops H. P., Flood J. and Amann R. (1996). In situ analysis of nitrifying bacteria in sewage treatment plants. Wat. Sci. Technol. 34: 237 – 244

Waller U. (2000). Tank culture – including raceways and recirculating systems. In K.D. Black (ed.) Environmental impacts of aquaculture. Sheffield Academy Press

Waller U., Schiller A., Orellana J. and Sander M. (2002). The growth of young seabass *Dicentrarchus labrax* in a new type of recirculation system. ICES Council Documents

Waller U., Sander M. and Orellana J. (2005). A "low energy" commercial scale recirculation system for marine finfish. European Aquaculture Society Special Publications 35: 459 – 460

Watson S. W., Valois F. W. and Waterbury J. B. (1981). The family Nitrobacteraceae. In: Starr M. P., Stolp H., Truper H. G., Balows A. and Schlegel H. G. (eds) The Prokaryotes. Springer-Verlag, Berlin, pp. 1005 – 1002

Wecker B. (2002). Anorganische Stoffflüsse in einem experimentellen Seewasserkreislauf mit definiertem Fischbesatz. Diplomathesis at the Institute for Marine Science Kiel, Germany. pp 99

Wecker B. (2006). Nährstofffluss in einer geschlossenen Kreislaufanlage mit integrierter Prozesswasserklärung über Algenfilter: Modell und Wirklichkeit. PhD-thesis at the Leibniz-Institute of Marine Science, Kiel, Germany. 157pp

Wecker B., Kube N., Bischoff A. A. and Waller U. (2006). MARE – Marine Artificial Recirculated Ecosystem: feasibility and modelling of a novel integrated recirculation system. In Kube N. (ed.) The integration of microalgae photobioreactors in recirculation systems for low water discharge mariculture. PhD-thesis at the Leibniz-Institute of Marine Science, Kiel, Germany. (2006)

Weirich C. R. and Riche M. (2006). Acute tolerance of juvenile Florida pompano, *Trachinotus carolinus* L., to ammonia and nitrite at various salinities. Aquacult. Res. 37: 855 – 861

Winberg F. G. G. and Duncan A. (eds.) Methods for the estimation of production of aquatic animals. Academy Press, London (1971)

Wijkstrom U. N. (2003). Short and long-term prospects for consumption of fish. Veterinary Research Communications 6/89: 48 – 52

Wilson R. P. and Moreau Y. (1996). Nutrient requirements of catfishes (Siluroidei). In Legendre M and Proteau J (eds.) Aquatic living resources, Gauthier-Villars, Paris, France (1996)

Wouters R., Zambrano B., Espin M., Calderon J., Lavens P. and Sorgeloos P. (2002). Experimental broodstock diets as partial fresh food substitutes in white shrimp *Litopenaeus vannamei* B. Aquacult. Nutr. 8: 249 - 256

www.biomar.dk
(Ref type: Electronic Citation; date 02.05.2006)

www.bran-luebbe.de
(Ref type: Electronic Citation; date 28.07.2006)

www.ika.de
(Ref type: Electronic Citation; date 22.03.2006)

www.vitfit.de/spurenelemente.htm
(Ref type: Electronic Citation; date 29.08.2006)

www.wtw.com
(Ref type: Electronic Citation; date 24.05.2006)

Y

Yingst J. Y. and Rhoads D. C. (1980). The role of bioturbation in the enhancement of bacterial growth rates in marine sediments. In: Tenore K. R. and Coull B. C. (eds.) Marine benthic dynamics. Univ. of South Carolina Press, Columbia pp. 407-421

Gefördert durch das
Stipendienprogramm der
Deutschen Bundesstiftung Umwelt

Deutsche Bundesstiftung Umwelt

i want morebooks!

Buy your books fast and straightforward online - at one of world's fastest growing online book stores! Environmentally sound due to Print-on-Demand technologies.

Buy your books online at
www.get-morebooks.com

Kaufen Sie Ihre Bücher schnell und unkompliziert online – auf einer der am schnellsten wachsenden Buchhandelsplattformen weltweit! Dank Print-On-Demand umwelt- und ressourcenschonend produziert.

Bücher schneller online kaufen
www.morebooks.de

VDM Verlagsservicegesellschaft mbH
Heinrich-Böcking-Str. 6-8   Telefon: +49 681 3720 174   info@vdm-vsg.de
D - 66121 Saarbrücken   Telefax: +49 681 3720 1749 www.vdm-vsg.de

Printed by Books on Demand GmbH, Norderstedt / Germany